William Gates LeDuc

Information in Relation to Disease Prevailing Among Swine and

Other Domestic Animals

William Gates LeDuc

Information in Relation to Disease Prevailing Among Swine and Other Domestic Animals

ISBN/EAN: 9783337242725

Printed in Europe, USA, Canada, Australia, Japan

Cover: Foto ©berggeist007 / pixelio.de

More available books at **www.hansebooks.com**

INFORMATION

IN RELATION TO

DISEASE PREVAILING AMONG SWINE

AND

OTHER DOMESTIC ANIMALS,

COMMUNICATED

TO THE SENATE BY THE PRESIDENT OF THE UNITED STATES, IN COMPLIANCE
WITH A RESOLUTION OF THE SENATE OF FEBRUARY 20, 1878.

WASHINGTON:
GOVERNMENT PRINTING OFFICE.
1878.

MESSAGE

PRESIDENT OF THE UNITED STATES,

COMMUNICATING,

In answer to a Senate resolution of February 20, 1878, information in relation to the disease prevailing among swine and other domestic animals.

FEBRUARY 23, 1878.—Read, ordered to lie on the table, and be printed.

To the Senate of the United States :

I transmit herewith, for the information of the Senate, the reply of the Commissioner of Agriculture to a resolution of the Senate, of the 20th instant, " relative to the disease prevailing among swine," &c.

R. B. HAYES.

EXECUTIVE MANSION, *February 27, 1878.*

DEPARTMENT OF AGRICULTURE,
Washington, D. C., February 26, 1878.

SIR : In compliance with a resolution of the Senate, adopted on the 20th instant, calling upon me for such information as may be in my possession relative to the disease prevailing among swine, commonly known as " hog cholera," with such suggestions as I may deem pertinent in this connection, I have the honor to herewith transmit a large number of letters, from almost every section of the country, relating to this,and the many diseases to which all other classes of domestic animals are subject. For some years past the local press, and especially the agricultural journals of the country, have been calling attention to the increase of diseases among farm-stock, and the consequent heavy losses annually sustained by our farmers and stock breeders and growers. I regarded the subject as one of such vast importance to the productive industries of our country, as to demand the immediate attention of this department, and early in the month of August last I caused the following letter to be addressed to each member of the House of Representatives :

DEPARTMENT OF AGRICULTURE,
Washington, D. C., August, 1877.

SIR : This department is desirous of making a thorough investigation into the causes of the many diseases now, and for some years past, prevailing with such fatal effect among the farm animals of this country. In order that this work may be facilitated, and the department put in possession of information that will enable it to form a correct understanding of the extent, nature, and character of these diseases, and of the remedies indicated therefor, I would thank you for the names and post-office address of some of the more prominent stock and poultry breeders and dealers of your district, and also the address of respectable veterinary surgeons, who have had experience in this direction. These names should include those of persons residing in localities where diseases of a general and local character prevail, either among horses, cattle, sheep, hogs, or poultry.

I remain yours, very respectfully,

WM. G. LE DUC,
Commissioner.

In compliance with the request contained in the foregoing letter, the department was promptly placed in possession of the names of a large number of prominent farmers, stock growers, and veterinary surgeons in the various Congressional districts of the country, to whom the following circular letter was addressed:

DEPARTMENT OF AGRICULTURE,
Washington, D. C., August, 1877.

SIR: This department being desirous of making an investigation into the causes of diseases now, and for some years past, prevailing among all classes of farm animals, desires your assistance and co-operation in the proper consideration and determination of a subject of such great importance to the stock breeders and growers of this country. With the view of saving millions of dollars annually by the timely use of such remedies and preventives as are now known or may be discovered by this investigation, the department desires as speedily as possible to be put in possession of such facts as may have come under your observation in relation to diseases affecting horses, cattle, sheep, hogs, and fowls, either in your own neighborhood or in adjacent localities. Have the kindness to give as complete a diagnosis of the disease as possible, stating the duration of the attack, its average fatality, what remedies, if any, are used, and with what success. In localities where diseases prevail among more than one class of animals, separate statements should be made under the name of the animals affected.

This information will be laid before Congress as the basis of securing an appropriation to meet the expenses of a thorough investigation of diseases of all classes of farm animals.

Very respectfully, &c.,

WM. G. LE DUC,
Commissioner.

The responses to this circular indicated such heavy losses, among swine, from the various fatal diseases affecting this class of animals, that I at once determined to secure, so far as the facilities within my reach would permit, as accurate returns as possible from the various States and Territories of the Union the number of hogs annually raised, what proportion of those suffer from disease, and what proportion of those affected die, the aggregate value of the annual losses among this class of animals, and also the aggregate value of annual losses among all other classes of domestic animals from the various diseases to which they are incident. In order to secure this information a copy of the following letter was addressed to one correspondent in each county of the United States:

DEPARTMENT OF AGRICULTURE,
Washington, D. C., December 21, 1877.

SIR: This department desires to obtain reliable information in regard to the losses of swine by cholera and all other diseases incident to this class of farm animals. We would, therefore, thank you for early and definite answers to the following questions:

1. What number of hogs are annually raised in your county ?
2. What proportion of hogs suffer from disease ?
3. What proportion of those attacked by disease die ?
4. What is the money-value annually lost by disease among swine in your county ?
5. What is the money-value of all other classes of farm animals annually lost by disease in your county ?

Very respectfully,

WM. G. LE DUC,
Commissioner.

Out of two thousand four hundred and forty-seven counties (the number composing all the States and Territories of the United States,) returns from one thousand one hundred and twenty-five counties have been received. These returns are still coming in slowly, and will, perhaps, not be fully reported for some weeks to come. The data, so far received, have been condensed into the following tabular statement:

Statement showing the number of swine annually raised in the United States, the number lost by various diseases and the value of such losses; also the value of all other classes of domesticated animals annually lost by disease.

Names of States and Territories.	Total number of counties.	Number of counties reported.	Number of hogs raised annually in State.	Number affected with various diseases.	Proportion of those attacked that die.	Money-value of losses.	Money-value of annual losses of all other classes of domestic animals.	Total value of annual losses of all classes of domestic animals.
Alabama	66	23	253, 250	51, 110	. 5877+	$88, 740	$184, 550	$273, 290
Arizona Territory	6	2	1, 850					
Arkansas	74	27	251, 176	62, 333	. 6750	142, 095	79, 400	221, 495
California	52	15	173, 339	3, 110	. 6057+	8, 175	58, 490	66, 665
Colorado	29	9	10, 635				23, 300	23, 300
Connecticut	8	7	74, 748	1, 270	. 0611+	10, 974	49, 870	60, 814
Dakota Territory	43	8	17, 500	150	. 4375	625	8, 898	9, 523
Delaware	3	2	20, 000	1, 000	. 7166+	4, 500	8, 000	12, 500
Florida	39	16	95, 909	18, 550	. 6096+	45, 605	75, 840	121, 445
Georgia	137	69	472, 631	62, 473	. 5878+	228, 496	300, 155	528, 651
Idaho Territory	10							
Illinois	102	65	2, 406, 449	491, 203	. 0763	1, 703, 327	593, 737	2, 297, 064
Indiana	92	49	1, 384, 032	326, 488	. 7248+	1, 446, 798	502, 700	1, 949, 498
Indian Territory	*9	*1						
Iowa	99	70	2, 055, 899	221, 158	. 6375	1, 884, 175	440, 165	2, 324, 340
Kansas	84	39	629, 296	70, 340	. 5259+	217, 165	208, 290	425, 455
Kentucky	116	30	581, 681	118, 163	. 6842+	412, 403	301, 487	713, 890
Louisiana	58	16	206, 542	12, 620	. 5250	33, 483	52, 700	86, 183
Maine	16	11	46, 767	730	. 4528+	4, 750	47, 532	52, 282
Maryland	23	14	202, 972	13, 595	. 6563+	55, 170	120, 525	175, 695
Massachusetts	14	3	27, 600	370	. 3900	2, 620	5, 850	8, 470
Michigan	77	43	401, 127	11, 013	. 4911+	49, 560	319, 625	369, 185
Minnesota	76	38	200, 426	2, 245	. 5758+	7, 985	100, 207	108, 192
Mississippi	75	34	316, 304	43, 015	. 6395+	112, 416	366, 665	479, 081
Missouri	114	63	2, 144, 084	437, 776	. 6692+	1, 351, 265	419, 264	1, 770, 929
Montana Territory	10	3	3, 300	25			1, 000	1, 000
Nebraska	65	25	234, 294	31, 160	. 6525	128, 925	82, 269	211, 194
Nevada	14	1	1, 000	150	. 9000	1, 300	4, 300	5, 600
New Hampshire	10	4	12, 771	275	. 4233+	2, 200	34, 100	36, 300
New Jersey	21	3	11, 000	240	. 8500	2, 170	16, 720	18, 890
New Mexico Territory	13	3	26, 000	1, 240	. 7000	2, 500	17, 000	19, 500
New York	60	26	270, 786	10, 387	. 3901+	32, 210	165, 934	198, 144
North Carolina	94	47	634, 048	140, 724	. 5580+	424, 825	190, 544	599, 369
Ohio	88	60	1, 687, 788	24, 890	. 5900+	503, 338	334, 608	837, 946
Oregon	23	8	28, 345			235	15, 500	15, 735
Pennsylvania	66	39	635, 387	23, 871	. 5916+	112, 999	155, 000	267, 999
Rhode Island	5	2	11, 848	30	. 7500	300		
South Carolina	32	15	133, 891	22, 915	. 6058+	60, 100	64, 300	124, 400
Tennessee	94	49	907, 606	175, 941	. 6244+	489, 515	311, 550	801, 065
Texas	154	40	597, 310	91, 979	. 5147+	235, 969	417, 631	653, 600
Utah Territory	20	7	17, 690	65	. 4166	360	23, 350	23, 710
Vermont	14	10	63, 591	3, 029	. 5237+	19, 035	42, 850	61, 885
Virginia	99	60	713, 275	59, 868	. 5104+	168, 174	217, 544	385, 724
Washington Territory	24	1	2, 600	30	. 2000	50	1, 000	1, 050
West Virginia	54	31	294, 533	35, 439	. 6 23+	66, 490	65, 500	131, 990
Wisconsin	60	39	671, 995	9, 622	. 5668+	26, 461	137, 995	164, 456
Wyoming Territory	5	2	67					
Total	2,447	1,125	18, 987, 342	2, 599, 542	. 5889+	10,091, 483	6,561, 945	16,653, 428

* Nations.

While I do not vouch for the accuracy of these returns, they are, perhaps, as reliable as can be procured, except by a systematic census of each county. In but few cases are the returns based upon the estimate of any one individual. In most cases these correspondents are conversant with the agricultural interests of the counties in which they reside, and in making up their returns they did so only after consultation with their neighbors and with the officers of agricultural societies and local granges, where such associations exist.

Our wide extent of country and its great diversity of temperature and variation of climate, the severity of frosts in some sections and the intensity of heat in other localities, render farm-stock liable to the attacks and ravages of almost every disease known in the history of domestic animals. So general and fatal have many of these maladies grown that stock breeding and rearing has, to some extent, become a precarious calling instead of the profitable business of former years. This would seem to be especially true as it relates to swine. Year by year new diseases, heretofore unknown in our country, make their appearance among this class of farm-animals, while older ones become permanently localized and much more fatal in their results. Farmers, as a general thing, are neglectful of their stock, and pay but little attention to sporadic cases of sickness among their flocks and herds. It is only when diseases become general, and consequently of an epidemic and contagious character, that active measures are taken for the relief of the animals afflicted. It is then generally too late, as remedies have ceased to have their usual beneficial effects, and the disease is only stayed when it has no more victims to prey upon.

This interest is too great to be longer neglected by the general government. Not only the health of its citizens, but one of the greatest sources of our wealth, demands that it should furnish the means for a most searching and thorough investigation into the causes of all diseases affecting live stock.

While a large number of the diseases to which farm-animals are subject are familiar to skilled veterinary surgeons, it should be borne in mind that but few sections of the country are blessed with professors in this science, as reference to the letters herewith transmitted will abundantly show. In most cases the stock-breeder himself assumes the important functions and responsibilities of the surgeon, and without the least knowledge of veterinary science he proceeds to diagnose the disease and dose the animal with a drug of which he is ignorant, or of the effects of which he is unable to judge until the animal has passed beyond the hope of recovery. Where such is the practice it is not strange that nine cases out of ten prove fatal.

It may be urged that works on veterinary surgery describe most of the diseases to which farm-animals are subject, and point out the best-known remedies for the same. Admit this, and yet the desired relief is not afforded thereby. These works are expensive and but few farmers can afford to purchase them; a still less number possess the qualifications necessary to comprehend them. What would then seem the wisest policy is that which may result from an investigation of the character proposed, and the results disseminated gratuitously through means of cheap annual reports from this department. By such means farmers and stock-breeders would soon become familiar with the general symptoms of the more prevalent diseases, and also be enabled to apply intelligently such remedies as science has or may point out.

Some very interesting investigations in this matter have been commenced, and are now being prosecuted with such vigor as is possible with the very insufficient appropriation that can be devoted to the purpose.

The correspondence herewith transmitted contains information which, if heeded by the farmers and stock-growers of the country, will result in a better sanitary condition, and a consequent diminution of disease among all classes of domesticated animals.

I have the honor to remain, very respectfully, your obedient servant,

WM. G. LE DUC,
Commissioner.

The PRESIDENT OF THE UNITED STATES.

DISEASES OF FARM-ANIMALS.

The following letters relating to the diseases of farm-animals have been received by the Commissioner of Agriculture:

Mr. JOHN BROOKS, of Princeton, Mass., writes as follows:

I last year raised fifteen Hereford steer calves. I bought twenty, and out of this number lost five by a disease called "blind staggers." I think eight or ten of them in all were affected; and I lost five before I discovered the nature of the disease and found a remedy. I do not know but there is a better remedy than the one I used. The calves, when first attacked, would not take nourishment, but held their heads up and walked around the pen until they were exhausted, and then, in about two or three days, would lie down and die. I lost five in that way. Four I saved in the following manner: I turned new milk down them three times a day—two quarts at a dose. I mixed about one half pint of castor-oil with the first dose every other morning. I kept this up about six days, when they again commenced to take nourishment. They appeared weak for about eight days after they commenced improving. They lost flesh, but not to any great extent, and seemed to winter as well as those that had not been sick.

The calves were constipated, and I gave the oil to remove this difficulty. I can define no reason for this sickness. I have lost calves for a number of years in the same way, but now think, if taken in season, they can be cured by the above remedy.

THEODORE S. VERY, veterinary surgeon, vice-president of the United States Veterinary Association, writes as follows from Boston, Mass.:

I regret that I cannot, from experience, relate facts about the contagious diseases of cattle, sheep, hogs, and fowls. Having resided always in this city, somewhat remote from farming and stock-raising districts, my practice has included for the greater part only the treatment of the diseases of the horse. Of these there are not a few concerning which a large amount of practical good would arise from a more thorough and positive establishment of the causes leading to their development and propagation.

The epizootic influenza of the fall of 1872, caused an immense aggregate loss by death, by loss of services while animals were sick, and in depreciation in values where the effects of the disease lessened the vitality of horses for a long time subsequent to its first attack. Possibly a thorough search for its causes might prevent a similar general outbreak, and inquiries having such an end in view should receive the attention, the support, and encouragement of the general government.

The disease known as the cerebro-spinal-meningitis occurs as an epidemic among horses, and is caused by a peculiar poison affecting the system in a specific manner, producing like symptoms—differing, of course, in degree—in all cases. Nothing is known concerning the exact nature of this poison, any more than of some others producing disease in a similar way. It causes great losses to horse-owners in seasons when it prevails, and has occurred extensively in certain localities, at different periods, for the past five years.

Glanders in horses—a most contagious, deadly, and incurable disease—has been quite prevalent in Boston and vicinity during the past five years. The poisonous particles of the disease are seldom entirely removed from stalls and stables where horses having the disease have lived. In my opinion this malady might, under certain conditions, become quite general. If it should, the danger therefrom would be incalculable. Stringent State laws should insist upon killing every animal so affected, and provide for the unmistakable removal of every trace of the disease from stalls and stables where it has existed, under the supervision of some qualified person.

A number of other diseases of the horse, the prevention of which is possible and of great importance to the public welfare, continue to exist.

Mr. F. M. HENDERSON, Leesburg, Loudoun County, Virginia, says:

Some six years ago Mr. J. T. Steadman, of this place, had a flock of fine fowls very much affected by chicken cholera, so called. I advised corn burnt on the cob thrown to them, and it acted immediately with wonderfully good effects. He not only lost no more fowls, but the disease soon disappeared.

I am told that hogs kept in pens and liberally supplied with charcoal very rarely have any disease, and I would certainly prefer the charcoal in the shape of burnt corn and cob, as being softer and possessing some real nutriment.

Mr. R. A. Steele, Lawrence, Kans., says :

In reply to inquiries in regard to diseases among farm-animals in this neighborhood, I would say that the most serious is a disease among hogs, commonly known as "hog cholera." In October, 1876, I had on hand seventy-five hogs, averaging 115 pounds per head, for the purpose of feeding with or following cattle. They were mostly of the Berkshire breed, and seemed in fine condition. The feed and water were good. In December they commenced coughing, and soon after dying, until I lost over half the number. I finally turned them out in a corn-field which contained some wet ground, in which they spent most of the time rooting. The disease was arrested, and no more of them died. They visited some of my neighbors' hogs, but did not convey the disease to them. I examined several of those that died, and came to the conclusion that the lungs were affected.

I found the same disease existing among hogs throughout the country. I do not regard it as the same disease of which so many hogs died in 1873 and 1874. I think they were affected with worms.

My opinion is that hogs are forced, and fed, and bred too young—a mushroom growth. As a remedy, we should use matured sows and males for breeding, and allow them to run in pastures. They should not be fed and fattened until a year or eighteen months old. To insure healthy meat and do credit to the hog product, such a system must be adopted.

Cattle as a rule, are healthy, but there is some complaint among calves. Those in good condition in the fall seem to be liable to the attack of a disease known as "black leg." My remedy has been a preventive (I have never cured one), as follows: Salt well, with a small quantity of saltpeter.

Mr. M. B. Hine, Austerlitz, Kent County, Michigan, says:

This portion of Michigan has thus far been comparatively exempt from any prevailing disease among our domestic animals, with the exception of epizootic among horses during the fall and winter of 1872–'73, which was here attended with but little loss by the death of the animals themselves, but the produce of the mares then in foal proved to be weak and debilitated. Since that time there has been considerable distemper occurring annually not unlike the epizootic, but quite mild in form, which readily yields to an outward application of some counter-irritant on the glands, at the same time keeping the bowels relaxed by feeding bran mashes, meanwhile working lightly.

Last spring there was a general fatality among the young pigs, and in some instances with the autumn pigs in the latter part of the winter, there being no apparent sickness discovered prior to their death. At least such was the case in this immediate vicinity ; but I noticed that this all disappeared as soon as the hogs were turned out to grass. The conclusion I arrived at is, we must furnish a greater variety of food for our hogs during our long winters, particularly of roots and vegetables, instead of feeding all corn, as is usually the practice with most western farmers.

Robert Vanvoorhis, importer and breeder of thoroughbred American merino sheep, Monongahela City, Pa., says :

I have been a breeder of American merino sheep for over thirty years. For six years or more the sheep of this vicinity have been afflicted with a disease commonly known as "paper skin." It has proved very fatal, especially to young sheep, thousands having died annually. It is more prevalent and more fatal to young sheep in August, about the time of weaning. Sheep of my own breeding have never been affected by the disease; but I have lost a great many lambs that I had purchased of others. Those attacked, if they did not die the first fall, were sure to do so the next season. I bought twenty-five ram lambs, which I took extra good care of the first year. Though they did not thrive as well as those of my own breeding, I had hopes that they would escape the disease. Last July I noticed that they began to show symptoms of the malady, and, having a large flock of yearlings, I took out those affected and gave them extra good care. I commenced to feed sulphur and copperas to them, but without any perceptible effect. After five had died I doubled the dose, giving a tablespoonful of pulverized copperas every other day ; but this did not seem to stay the ravages of the disease. I continued this until I had but five left. The dung of those that died last was white with worms, which were from one to four inches in length. After my entire flock of twenty-five had died, I thought if I had commenced with a heavy dose as soon as the first symptoms were observed, I might have saved some of the lambs. In order, therefore, to test the remedy further, I informed some of my neighbors, who gave it a thorough trial, but without success.

I was recently in the eastern part of Ohio, where I found the sheep affected with the same disease. It seems to be as fatal there as it is here. I am in receipt of a letter from Col. J. W. Watts, of Martin's Depot, S. C., who informs me that he has lost a great many sheep from the same disease. I last November visited Vermont, where I also found the same disease prevailing to an alarming extent. The diseased sheep do not lose flesh. They seem to lose blood, however, for in a short time their skin becomes perfectly white. Their eyes also become white, the ears droop, and they are apparently much exhausted. They drink water freely, eat salt, gnaw at boards, and take up whole mouthfuls of dirt, but eat neither grass nor hay. They seem full and in good condition up to the time of their death. When opened no blood is found in their veins, but the stomach and intestines are full of worms, which have collected in bunches or knots.

Various remedies have been tried here, but without any perceptible effect. If the disease is not soon checked many large dealers and breeders will lose entire flocks of valuable animals.

Hon. HARRIS LEWIS, Frankfort, N. Y., says:

In reply to your favor of the 31st ultimo, I would say that this county (Herkimer) is almost wholly devoted to dairying, and that the only diseases of farm stock affecting us are those which affect dairy-cows, as but little other farm-stock is kept. Epidemic abortion has prevailed here among our herds more or less for the last eighteen or twenty years, and a part of this time to an alarming extent in many herds, as high as 90 per cent. of the cows aborting.

At the annual meeting of the American Dairymen's Association, held at Rome, N. Y., in January, 1876, a committee of three was appointed (of which I was one) to petition Congress to offer a reward of $10,000 for the discovery of the cause and a remedy for abortion. But the committee appointed, believing this sum ten times too small, never took any action in regard to the matter, and hence it has rested ever since. But if the dairymen of the United States can rely upon the Commissioner of Agriculture to aid them in the work of investigation, we will take new courage and see what can be done with his aid to relieve dairymen and stock-breeders of this terrible scourge, by which more than a million dollars has been lost each year for the past fifteen years

Mr. ELI AVERY, Clayville, Oneida County, New York, says:

Seven years ago I lost seven swine. Nothing unusual in their condition was noticed. They were fed in the morning, all eating well, but at noon feeding one was found dead and bloated, his legs standing out stiff like the legs of a bench. The others fed well, but at night two more were found in the same condition. This continued until seven had died. They turned purple as soon as they died. Others were lost in our town at that time from the same disease. Mine were principally fed on skimmed milk.

I have had occasionally, in wet weather, horned cattle affected with hoof-ail, or foul foot, as the farmers call it. This disease is very easily cured by cleaning the hoof with acid and covering it with tar.

F. D. RUICK, La Grange, Ind., says:

During the prevalence of chicken cholera in this section it is very fatal. Chickens attacked with it will sometimes live a day or two, but generally they will die within a few hours. I have fed a hundred head in the morning, all apparently in good health, and at noon have found half of them dead, and perhaps half of those remaining were staggering around like so many drunken men. The disease is no doubt contagious, and if the chickens affected are not at once separated from the well ones the entire flock will soon be inoculated.

This year I have lost but three chickens from the malady. As soon as I discovered that they were affected I separated them, giving the well ones a fresh coop, and fed them freely with red pepper (capsicum) and sulphur. The result was that I saved the balance of my flock.

The feathers on the breasts of some of the chickens, when first attacked by this disease, become ruffled; the breast hangs down between the legs and appears to be full of water, like one afflicted with the dropsy.

J. B. BLOOMER, V. S., Wauconda, Lake County, Illinois, says:

Horses here suffered severely with epizootic distemper, and for two years thereafter were disposed to severe influenza. The epizootic left the mucous membrane so much inflamed that a slight cold would seriously affect them; but this would generally yield readily to simple remedies.

With the exception of milk-fever, cows are unusually healthy. For this disease, I am sorry to say, I have found no sure remedy. On an average, I have lost about one-third of the cows that I have treated. However, in no instance have I lost a case where I have been called upon in the early stages of the disease. My practice is:

Counter-irritants of mustard on the back, with very hot cloths changed every five minutes. Medicines: Digitalis, tartar-emetic, and niter. Diagnosis: Total loss of strength in their feet within three hours after the attack; high fever, constant moaning, legs cold and sprawled out.

Our sheep suffered throughout last winter with a disease similar to the epizootic. In many cases all the yearlings in some flocks died. The older ones were not affect-d. The attack would commence with hard coughing, loss of appetite, general debility, &c. The farmers pronounced it fatal. When told what to do to relieve them they answered it was no use, as medicine had failed to have any effect. Should the disease occur again during the coming winter I will give it close attention and report.

Mr. NATHANIEL VOSE, Whittier, Ill., says:

Last spring, horses here were attacked by what is commonly known as "horse-distemper," with some difficulty of breathing, &c. Their heads swelled to the point of the muzzle, and sores commenced to gather and break on all parts of the head, and discharged freely, with the usual running at the nose. The usual remedies of physic and smoking condition-powders with leather, and in some cases roweling and bleeding, were resorted to, but they were of no avail, as death ensued after one or two months. From my observation of a yearling colt, it seemed to be affected like a person with the scarlet fever, excepting there was no difficulty in swallowing food or drink. The gatherings continued and the colt became very much emaciated, yet was able to walk about until it died. Two and three year olds have died with it, but no old mares. It seemed to be a malignant type of horse-distemper. The disease was similar to others which had the distemper very light, only the head swelled enormously in the fatal cases.

The foot-rot in sheep has heretofore prevailed to some extent, but is about eradicated. Sulphuric acid and copperas are the general and successful remedies.

Last year there were some losses by hog cholera, but no cases came under my personal observation. This stock is generally healthy hereabouts. A short time ago I noticed one of my pigs had what, I believe, is called the "thumps." This was the only case I ever saw. The pig would pant or jerk almost like a person with the hiccoughs, only the jerk seemed more in the abdomen than in the chest. It grew thin and died after about six weeks.

A. M. DICKIE, M. D., Doylestown, Pa., says:

I have given some little attention to the ailments of fowls, as I keep a few and am interested in them. The diseases incident to the poultry-yard are very little understood, and the result is enormous aggregate losses every year. In the general investigation and study of the diseases of farm-stock, veterinary science has, so far, ignored or omitted to study the ailments of poultry, mainly, perhaps, from the fact that poultry are looked upon as inferior, or subsidiary farm-stock, and of little or no account anyhow. This is a misapprehension, because the poultry interest is really an important one, susceptible of almost indefinite expansion and usefulness.

The hinderances to poultry-keeping may be arranged in three classes: 1. Parasitic diseases. 2. Catarrhal diseases. 3. Poultry cholera.

In the first of these classes, the main trouble is the *gape* disease, produced by a parasitic worm in the trachea of young chickens and turkeys.

In the second class the principal disease is *roup*. This is common to all ages, and prevails mostly from November to May, and is much more prevalent north of the fortieth parallel than south of it. I think there are at least four distinct diseases included in the general term *roup*.

The prevailing disease in poultry in the summer months, or from May to November, is poultry cholera. This is much the most serious and fatal of the hinderances, and has discouraged many poultry-keepers south of the fortieth parallel, as it is mainly developed south of this line.

These diseases are all epidemic in character. They extend over large sections, are very destructive, and in some localities have greatly discouraged people to the detriment of the interest in poultry production.

If poultry-keepers knew how to manage or control these three classes of ailments, most of the hinderances would be overcome. Any investigation which will tend to remove or overcome them will be gladly accepted by the people as a needed assistance in protecting an important and growing industry.

The general public has no appreciation of the importance and value of the poultry industry in our country, and especially on the North Atlantic slope and Lower Lake region. The annual products of the poultry-yards of the nation are variously estimated at from $200,000,000 to $450,000,000. The truth is probably between these extreme figures.

The annual poultry products of Bucks County are very near $2,000,000, and amount to more in value than any other industry pursued in the county as a specialty. It leads the dairy interest, the grain interest, the fruit interest, and all other kinds of live stock put together. But this is probably the leading county in the Union in poultry production. There is nothing to hinder the business being generally extended, except the drawback resulting from ignorance of the management and the prevailing diseases. If your department can secure an appropriation to conduct an inquiry into the diseases of domestic animals, fowls included, it would certainly be to the advantage of our agriculture.

Mr. S. E. STOWE, Grafton, Mass., says:

Abortion in cows, here in the central part of Worcester County, is becoming quite an alarming disease. Since receiving your letter, by inquiry I find that about one-fourth of our cows lose their calves, some at four, but the majority at from six to eight months along, causing a loss to the farmers of one-half their value for dairy purposes. There has been nothing done in this vicinity, or in the State, to find out the cause or discover a remedy. My own conclusion is, that the disease is caused by some weed that is eaten by the cows, both in grazing and in the cured hay.

Mr. J. S. DUNCAN, Cross Creek Village, Pa., says:

In reply to your letter I will say that we have various diseases among our farm-stock. The first is among cattle. The prevailing disease is among cows, commonly called milk-fever. The symptoms are as follows: About twelve hours after the cow drops her calf she becomes restless, switches her tail and moans occasionally as if in pain. The treatment is usually to physic the animal, bathe the back with mustard and warm water, and give ginger internally. The average loss of animals is about three-fourths of all attacked.

The prevailing disease among sheep is foot-rot. It commences between the toes. It first appears like a scald, and spreads, until finally the entire foot becomes affected. The remedies are, butter of antimony, blue vitriol, and nitric acid. Various other remedies have been tried, but without effecting a perfect cure.

The disease most common among poultry is cholera. The first sympton is extreme dullness. As a general thing they will sit on the roost until they drop off dead. They are usually sick from twenty-four to thirty-six hours. When this disease enters a flock it is so contagious and fatal that but few, if any, escape. Various remedies have been tried, but none have proved effectual.

Mr. J. B. KENDRICK, Monticello, Ky., says:

We suffer here mostly from hog cholera. It is impossible to give any diagnosis of the disease, for it manifests itself in numerous ways. I have seen them linger for months and ultimately recover, while others would die very suddenly. Again I have seen them in apparent good health, and, while eating, suddenly jump up, squeal, and fall over dead. Hogs turned on the mast or acorns last winter did very well apparently, but when killed many of them were found to be affected with worms. In numerous cases worms an inch or more in length had penetrated the heart and bowels. Cholera has generally been worse among our hogs after a good mast-year than any other time.

Chickens are also affected with what is known here and elsewhere as cholera. The part affected most seems to be the liver, which enlarges to two or three times its natural size. The fowl is always fat when attacked. The disease is very fatal. Guinea-fowls and geese are occasionally attacked.

Horses are comparatively free from diseases, except such as are brought on from bad treatment. I have known two or three mares unable to bring forth their foals, for which no cause could be assigned by their owners. * * * What produces the water-bag in colts after they are castrated? Several that I have castrated have been so affected, but I can assign no cause for it. It greatly injures the sale of young horses in the South.

Mr. LE GRAND BYINGTON, Iowa City, Iowa, says:

In my thirty years' experience as a farmer no subject has worried me so much as the "diseases prevailing with such fatal effect among my farm-animals," and upon no subject, let me add, is so lamentable ignorance prevailing, among those of my occupation, as upon the causes, preventives, and treatment of these prevailing diseases. In all that time I do not remember an instance of an animal of mine, of the cattle or swine species, that ever recovered from a serious illness. If you succeed in disseminating valuable information on this important matter, you will be remembered with gratitude; and in the effort you can rely upon my co-operation.

Mr. J. B. REID, Macon, Tenn., says:

I will say here that no one in this locality pretends to know the causes of the various diseases which from time to time afflict our farm-stock. Intelligent planters seem to think there is no cure for any of them, and it would seem true, as they invariably prove fatal. Within a radius of a half mile of this little village twenty head of milch cows have died in the last three or four months. The disease proved fatal in every instance. The government will act with wisdom in making an effort to stay these diseases, as the loss annually is immense.

Dr. J. R. WOOLFOLK, writing from the same place, says:

In answer to your inquiry, "What is the nature of the disease which now prevails among the hogs of your community?" I must say that from the very casual examination I made of one that died of the disease, I am not prepared to give a satisfactory diagnosis, and will only say that I found complete engorgement of the liver, with enlargement of the same. The lungs presented no indications of disease, nor did the intestines. There was a collection of serum or bloody water in the pericardium or investing membrane of the heart. There were no indications whatever of inflammatory action on any of the abdominal or thoracic viscera, except the liver. There was positive passive congestion of the capillary circulation generally, which I have reason to believe was not stasis anæmia, as might reasonably be supposed. It is the same disease which prevailed to such a destructive extent among the hogs in this country in the year 1868. From what I can learn the same may be said in relation to the disease among cows. There is an appearance of congestion of the capillaries; they are also infested with ticks. Some of the farmers say it is bloody murrain, while others believe it to be dry murrain.

Mr. H. SEVISON, Constantine, Mich., says:

We have had no diseases among horses, cattle, or sheep for several years past, but our hogs have been seriously affected with what is generally known as cholera. The disease has been a very peculiar one here. Some were affected in their hind limbs, others in their fore legs; some died very suddenly, while others would linger for months, and, after becoming mere skeletons, would lie down and die. The loss has been very heavy. No cause for the disease has as yet been discovered or remedy found. Some have thought that pure, clear water would prove a preventive, but such is not the case.

Mr. T. J. McDANIEL, breeder of standard and fancy poultry, Hollis Centre, Me., says:

The greatest drawback we have in poultry-raising in New England is roup (*Cynanche trachealis*), and canker or catarrh (*Czæna*). The former is characterized by a difficulty of respiration, particularly at each inspiration, while the expiration is less difficult. The fowl will raise and extend its head at each breath, thereby inducing coughing. This is sensibly increased at night, and will end in suppuration, during which stage it is highly contagious. Fowls so afflicted should be immediately killed or isolated from all others.

The causes of roup are insufficient ventilation or damp roosting-places; food that will induce catharsis, such as potatoes, sour milk, particularly buttermilk, overfeeding with fresh meat, &c. Last fall I got out of corn, and for three days fed boiled potatoes with a little meal; getting out of meal, I fed potatoes alone for three more days. At the end of this time we were visited by a storm of snow and sleet, and nearly every one of my fowls took cold. Several of them choked to death. Finally breathing with most of them became easier, when a purulent offensive discharge became established at the nostrils, and from the mucous membrane of the throat fauces in particular. Knowing the disease had become highly contagious at this stage, I at once separated them. I observed among those I had bred to standard (for fancy points exclusively) that the disease proved fatal in far the greater number of cases. For instance, among my brown Leghorns that were bred for exhibition purposes only, one cockerel had red ear-lobes—as nature designed—and he alone escaped, though confined with the worst cases, which were among and included nearly all the fine-bred birds.

In-breeding is another cause of failure in rearing pure-bred fowls. With common or native breeds it proves less disastrous, though it should never be continued for any length of time with these, as stamina is thereby decreased and the fowls rendered more susceptible of disease. The old adage holds good in the case of roup especially, that "an ounce of prevention is worth a pound of cure." However, my advice is to separate the fowls immediately upon discovering the least symptom, such as coughing, loud breathing, or a "wet beak." Fowls become contaminated much sooner if fed dough, as the dough that adheres to the beak will be picked off by others. The disease will also be communicated by drinking from the same vessel. The germs of the disease will float on the water and soon infect all.

I never knew of a case of roup in a flock that roosted in trees. If a flock (part of whose numbers have the roup) are at liberty, and are fed with corn scattered on the ground, not in the immediate vicinity of their pens, and should drink from a running stream of water, there need be but little to fear even if some of them should contract the disease; yet the affected cases should be separated, their roosting places thoroughly whitewashed, and all excrement removed. The fumes of burning tar during the night prove quite efficacious, if persisted in three nights in succession. Sulphate of iron (an ounce to one gallon of drinking-water) is the best remedy.

Fowls are most susceptible to diseases during the molting-season, or when the first snow-storms occur. Roup will soon be brought on by roosting in low and damp apartments during the winter months. Farmers who allow their fowls to roost high in their barns are seldom troubled by this malady.

Sulphur and lard rubbed on the heads of young chicks for the purpose of killing lice, though effective in destroying this pest, will soon bring on roup—sore eyes especially. I have had a dozen little chicks moping around with their eyes closed, and if they had not been fed by hand would soon have died of starvation. If roosting-houses become infested with lice, whitewash is the sovereign remedy, for a flock of poultry covered by these pests will, sooner or later, take roup and its concomitant troubles.

Mr. RALPH W. MILLS, Webster Groves, Mo., breeder of poultry, says:

My experience with fowls extends through a period of eight consecutive years, prefaced by a familiarity with this portion of the feathered race during boyhood. I have bred successfully, and in the order named, the varieties classed as games—White Crested, White Polish, Light Brahmas, Buff Cochins, White Booted Bantams, Gold Laced Sebrights, Partridge Cochins, Black Breasted Red Game Bantams, Plymouth Rocks, and Silver Spangled Hamburgs—all being of the kind popularly termed "fancy fowls."

As regards diseases affecting fowls coming under my observation, they are chiefly two in number, viz., cholera and roup; and what may prove as great a scourge as either, the plague of lice.

Cholera is in its symptoms not unlike the disease similarly named in the human disease. It is first observed in the character of the droppings, green in color, growing thinner, clearer, and more liquid with each subsequent evacuation, until, utterly weakened and prostrate, in a course of from twelve to forty-eight hours' duration, the fowl succumbs to death. During the attack great thirst is manifested, but indifference to food. I have been unable to learn that any person has ever positively determined the cause of this disease. My own opinion is that it is a generated poison (atmospheric), not unlike malaria, and dependent for its development upon certain favoring conditions in certain localities at certain seasons. It is contagious in some degree; and fowls having the disease should be promptly separated from those not affected, and those dying of it should be carefully buried at once, or burnt with brush or litter, to obviate the danger of infection.

I have but little faith and have had but indifferent success in "doctoring" the disease with anything in the nature of drugs given in doses. Four cases in five will result fatally. Dry, warm, clean, well-ventilated quarters, other than those lately occupied by the sick fowls, a complete change in the food offered and in the order of feeding, freely incorporating ground red or black pepper in all soft food given, with the "Douglass mixture" put in all the water placed before them to drink, will accomplish, together with an occasional disinfection of their premises by the use of carbolic acid in solution, and fumigation of their houses with roll brimstone and rosin placed on live coals, about all that can be done to cure and eradicate the disease.

Reference is made in this connection to the "Douglass mixture," a tonic in general use among experienced poultrymen everywhere, the formula of which originated with Mr. John Douglass, of the Walesley Aviaries, England. It is as follows: 1 pound sulphate of iron, 1 ounce sulphuric acid, 1 gallon water. Give a teaspoonful in each pint of water placed before the fowls to drink occasionally in health as a preventive; frequently in disease as a corrective. It is inexpensive and very efficacious. Upon the reasonable hypothesis that "prevention is better than cure," the suggestion is made that so far as they can be known, the *wants* of fowls should be supplied in order to keep them in health. Gravel, lime, grass, vegetable food, insects or animal food, liberty, fresh clean water, regularity in feeding, &c., are all essential to the healthfulness of domestic poultry.

The disease second in order, viz., roup, is well known, and, in its incipiency, can be successfully treated. It is the result of a cold, attacking the head; is similar to nasal catarrh in the human species. The disease arises from exposure to uneven and unwholesome temperatures, especially as maintained in the fowl-houses. Dampness here, want of light and ventilation, draughts of air, &c., are fruitful causes of its appearance and favorable to its perpetuation. The symptoms of an attack are, first, a thin, clear, mucous discharge from one or both nostrils, sneezing, and froth in the corners of the eyes. This froth can be seen to bubble when the fowl breathes. As the disease progresses (which it will certainly do if neglected), the discharge becomes more profuse, and changes in color and consistency, becoming decidedly yellow and thick, and eventually putrid—offensive to the sight and to the smell. The whole head becomes involved; the parts swell and become inflamed; the eyes close, and the patient, constantly falling off in condition, becomes helpless and unable to supply its wants, and finally dies. The disease may result fatally in two weeks, and may continue three months. I have cured two obstinate cases; in one of these, however, one eye was entirely lost. I once checked the disease that had attacked at least twenty of my fowls at one time by the timely use of vigorous sanitary measures alone—a thorough cleaning; fumigating (for like cholera it is contagious); changing of feed, &c.; the use of the Douglass mixture in the drinking-water, and a thorough cleansing of the parts affected in this disease—eyes, nostrils, mouth, and face—with "Lavrabaque's solution" of chlorinated soda (to be had of any druggist), diluted with an equal part of tepid water, applied with a small piece of sponge. In the instance referred to one application sufficed. In cases more advanced two applications per day, morning and evening, should be made until improvement follows. Warm water and vinegar, in equal parts, are useful cleansing agents instead of the solution mentioned. I have no confidence in the use of internal medicines in this or other diseases affecting poultry, except the "tonic" before mentioned. Change in the fowls' living-quarters, extreme cleanliness, disinfection and fumigation, are the general agencies that are to be employed in disease, so far as my observation goes.

As to the "plague" alluded to in the beginning of my letter (that of lice), in the language of a brother poultryman, "they are simply a disgrace," and perhaps after all, when allowed through neglect to multiply *ad libitum*, become the greatest of all obstacles in the way of successful poultry-raising. It may be set down as a rule that *fowls are never thrifty when infested with lice*. They are out of condition, and therefore especially liable to any of the diseases which infect their species. Prevention in this, as in the case of disease, is better than cure. Clean premises, dust-baths to wallow in; flowers of sulphur in the litter composing the nests; saturation of roosting-poles, or perches, with coal-oil; fumigation; application of hot whitewash to all parts of the fowl-houses, are effectual preventives of this scourge.

If the vermin have already obtained a lodgment upon fowls and in henneries, the same measures much more vigorously employed, in addition to those suggesting themselves as serviceable in improving the general condition of the flock, with the use of flowers of sulphur (a tablespoonful to a quart), in all the soft feed given them for a few days, will banish and destroy the nuisance.

This pest, of which I have definite knowledge, is of two varieties; a large kind, a sixteenth of an inch and over in length, quite in appearance like the genus that attacks squalid and untidy children, not very numerous on a single fowl, but leaving old fowls to prey upon young ones as fast as they appear. This kind is very damaging to little chicks, usually fastening upon the poll, and around the vent, and under the wings. Grease will kill them.

The other variety is far more troublesome in a general way, by reason of their great numbers, swarming in myriads in the places occupied by the poultry, and in places contiguous, literally overrunning the fowls, and almost deterring the keeper from entering the premises devoted to them, for they will get upon one's hands and clothes and are so infinitesimally small that they are with difficulty got rid of. This kind will drive a setting hen from her nest, and cause all the fowls to dread their quarters. They multiply to this extent *through neglect*. The remedy has been suggested.

It may be remarked in concluding, that the care, management, and treatment of fowls, in health and in disease, are essentially the same in the case of the choice, pure breeds in the yards of the "fancier" and mongrels produced by any sort of cross between varieties that are found in numbers upon the farmer's premises. In either instance, any measure adopted with a view to supplying the natural requirements of the creatures will be the most effectual means of improving their condition and enabling them to ward off disease, which in most cases, in my opinion, results from neglect.

GEORGE Y. PARRY, V. S., Newtown, Bucks County, Pa., says:

Typhus or Texas fever and pluro-pneumonia in cattle are the only diseases that exist in this section of Pennsylvania that we have any trouble in managing. The usual diseases have prevailed for the past few years that are common to horses and cattle, too numerous to write out a diagnosis.

If anything can be done by Congress to wipe out the two first-named diseases, I will be glad to assist in any way to accomplish it.

Mr. H. SHACKLEFORD, Woodbury, Tenn., says:

I have to remark that no disease of a fatal character exists among any of our farm animals except hogs. The disease generally known as hog-cholera has been prevailing in this county and other counties contiguous to it to an alarming extent, many of the farmers in the community having lost nearly their entire stock, so fatal has been its ravages in many localities.

As to a diagnosis of the disease, nothing definite or satisfactory has been arrived at so far as my investigation or information extends. The disease is developed in the same herd of hogs in various forms. For example: Some among the herd, apparently in good health, will be suddenly attacked with vomiting and purging, and will die in from twenty-four to thirty-six hours. With others the disease will assume a different form of attack. Some will make constant efforts to disgorge the contents of the stomach, which seems to be locked up in their bowels. This constipation continues with many of them from the time they first take the disease until they die. Many of them thus afflicted will live from one to two weeks, and I have known a few to wear out the disease and recover; but the few that survive rarely ever make thrifty hogs. I may further state that a great many hogs of different ages and sizes sicken and die without exhibiting the symptoms above pointed out. In a majority of cases which have come under my notice within the last two years, the disease can be easily detected in any herd of hogs by the symptoms indicated, apart from the vomiting, purging, &c. Whenever a farmer discovers among his hogs any that move around as though they were too lazy to get out of each other's way, afflicted with a squeaking cough, stiff in their joints, and when standing or walking hang their heads near the ground, with a most offensive effluvia exuding from their mouths and nostrils, accompanied by loss of appetite but insatiable thirst; also, manifesting a strong desire to find a warm place in which to lie down, and, when lying down, lie on their bellies instead of on their sides, they should be at once separated from the well ones; or, perhaps, the owner would be no worse off in the end to kill all such to prevent others from taking the disease from them.

I am of opinion that it will be needless for me to write anything on the subject of preventives and cures, as all the remedies heretofore introduced and tried in this section of country have been pronounced a failure by the most of those who have tried them. I am well satisfied in my own mind that not one of the many remedies which have been introduced and vouched for can be relied on as a cure, from the fact that what is commonly termed hog-cholera I believe to be a variety of diseases, and it is just as absurd to suppose that one remedy will cure all the diseases of hogs as that one remedy will cure all the diseases of man. Nevertheless, of the many remedies which have been brought to public notice, I doubt not but much good has been done by at least preventing disease, if not in some abstract cases effecting a cure.

I am an old man, and a firm believer in that old trite maxim, "An ounce of preventive is worth a pound of cure." If, therefore, the Department of Agriculture can, by further investigating the subject, discover a remedy which will check up or effectually stop a disease whose ravages hitherto have not been confined to any locality or climate, it will confer a lasting benefit on our nation.

Mr. JAMES M. MAYO, Whitaker's, Nash County, North Carolina, says:

In response to your circular letter of the 10th instant, I report as follows:

Horses.—One-half of 1 per cent. are subject to what is known among the planters as "Staggers." The animal seems sluggish and sleepy, eyes dull and sunken, ears cold, and pulse quickened. This continues from two to four days. The animal, at intervals, suddenly starts and walks, or rather staggers around in a circle, with head down. Of those affected, 99 per cent. die. It is noticed that there is more of this disease when we have a rainy spring than when the reverse is the case. The writer has cured one case. I drenched the animal with a solution or decoction of red pepper and salt once each day, and cut the forehead about two inches above and between the eyes, then running the blade of the knife down and up loosened the skin, thereby getting up a counter-irritant. I know of another horse cured by a similar process. I think this disease is due in a great measure to defective forage, bad and early grazing, when the animals are not accustomed to it—in short, when the planter, in anticipation of a short crop, desires to economize in feed and stints his animal. In 1867, we had an unusual amount of rain and bad crops, and the death rate by staggers was fearful. In Hyde County the rain-fall has been very great, and hence [crop prospects exceedingly poor, and the fatality this season has been much heavier than usual, as doubtless you have seen from the reports from that county. I state this much to show that with judicious management this fatal disease might be avoided. We also have the snuffling epizootic, that " comes on the wings of the wind." I use, and have seen used with good effect, carbolic acid, pine-tar, and other disinfectants. We have, in isolated cases, other dis-cases as cited in the books, and used the remedies as suggested, with the ordinary per centage of failure and success.

Cattle.—We can almost say our cattle are free from disease. We give them no care at all, even in winter, and what few die is the result of old age or starvation. The same may be said as regards sheep. There are but few in the county, and they take care of themselves.

Hogs.—Cholera, as we know it, is the disease that promises to make the raising of pork difficult in this country. The animal is taken with vomiting and running off of the bowels, no disposition to eat, general languor and listlessness. Of the old (one year and over) affected, 35 per cent. die; of the young pigs and shoats, 95 per cent. die. I do not think that the disease, once in the system of the boar or sow, ever leaves it entirely; for upon the sow having pigs again she will either have very few, or what she does have will soon die with this cholera.

Remedies.—1st. Put a small quantity of spirits of turpentine on the corn or in their feed three times each week or oftener, as it will do no harm. 2d. Feed all the slops and swill-feed you can, in which put saltpeter, red pepper, and salt. 3d. Keep salt at all times where the hogs can get all they want, and, by the way, keep it where all the animals can get at it at all times.

This dreadful disease was almost unknown in the days of our forefathers, and I have almost arrived at the conclusion that the raising of cotton has bred it. It is known that the eating of cotton-seed by hogs while the seed are in the process of fermentation will certainly kill them; and this, in my opinion, has brought about the disease. But the question is asked, "How does the cholera get up in Iowa and the Northern States, where they raise no cotton?" They buy the oil-cake, which is made of cotton-seed, and feed it to their hogs. A small percentage may die from eating poisonous mushrooms, but I do not believe that many die from that cause. On one of my plantations, on which I have raised a great number of hogs, I never knew a case of cholera, or any disease, until this year, when I lost between fifty and sixty pigs. The reason was this: My superintendent had made a compost-heap in which he had put a large percentage of cotton-seed, and the hogs had free access to it. So soon as the seed commenced to rot, the hogs eating them were taken with the cholera. If the farmers of the North will keep their hogs from cotton-seed and oil-cake, my word for it, they will never be troubled with cholera.

Fowls.—We denominate the main disease with them here "cholera." The fowl droops for a short time, and then commence spasms, from which they soon die. They are found dead under the roost and about the yard. I think this disease is partially due to inattention. The loss from cholera is about 5 per cent. For a remedy, feed them on small grain in moderate quantities. Mix in dough and feed once a week, or as the flock seems to need it, alum, red pepper, onions, and copperas. Keep marl or carbonate of lime where they can get it, and they will eat as they need it. Turkeys, ducks, geese, and peacocks are quite healthy, and I never knew one diseased.

Mr. GEORGE S. SELVIDGE, Wheatland, Mo. says:

Before proceeding to answer the interrogatories contained in your favor of the 24th ultimo, permit me to urge upon Congress, through your department, the absolute necessity of making such appropriations as will be required to meet the expenses of an investigation into the causes of the various diseases affecting farm-animals.

Horses are very healthy in this locality; no epidemic or contagion since the epizootic in 1872.

Since the passage of State laws prohibiting the grazing of Texas cattle on the prairies, mature cattle have been healthy. Calves are often affected with a disease known as "black-leg." About 90 per cent. of those attacked die. The symptoms are lameness in one leg (more generally, I believe, the right forward leg), ears pitched forward, and nose dry. This condition lasts from ten to thirty-six hours, when one in ten will probably begin to recover; the other nine, of course, die. I have tried blood-letting and active cathartics without effect. I have not known any other remedies tried. A rather remarkable feature of this disease is that the fat, well-fed calves are generally the first attacked; and if any escape, it is the lean ones. This feature of the disease has led some breeders to adopt as a preventive a seaton passed through the loose skin on the under side of the neck, by which a slight suppuration is kept up. Those who have tried it claim that this is an absolute preventive.

Sheep are healthy here, with the exception of an occasional case of scab or foot-rot, diseases too well known to require mention.

Under the head of diseases among hogs many pages might be written. In 1876 our farmers lost heavily, probably one-fourth of what should have been their income for the year. Some lost all, after feeding out the enormous crop of 1875.

The disease is what is generally known as hog-cholera. It presents itself in three distinct forms: One in vomiting and purging, presenting something of the symptoms of cholera in the human system. In another, severe constipation; and in yet another, the symptoms of quinsy, without the swelling under the throat. And yet we call it all cholera!

It would be hard to enumerate half the medicines that have been tried and found wanting as remedies for this disease. But after all we know absolutely nothing of the seat of the disease or its causes. Those who have not "doctored" at all have fared as well as those who have.

The same remarks are applicable as to fowls; we have suffered much and learned nothing. I have talked with a number of our best informed stock-men, and even with physicians, and find their theories differ so widely that I hardly think it worth while to give them.

Mr. W. B. FLIPPEN, Yellville, Ark., says:

Our horses are afflicted with no other diseases than distemper and blind-staggers. Farmers attribute the latter to feeding new ground corn or late corn, which is generally worm-eaten. Others attribute it to the horses eating unsound corn and worm-dust. I know of no certain remedy.

Cattle are occasionally affected with murrain, and what is here called black tongue. The tongue becomes red and raw on the upper side, and if not attended to promptly, turns brown or black on top, cracks open, and becomes so sore that the animal cannot feed, and in a short time will die. This disease prevails only at intervals. It is easily cured by washing the tongue two or three times with a solution of salt and copperas. I suppose the copperas alone would effect a cure, as those herds are never affected with it where copperas is mixed with their salt in the summer and fall months.

Cattle are also occasionally affected with what is here called Texas fever, but it is not incident to this locality, and prevails only along the line or route where droves of cattle from Texas have passed. The disease is very fatal, and I know of no remedy.

I have observed in this locality, twice within a period of forty years, a disease called "mad itch." I have never known a case cured. At first the animal appears feverish and not inclined to feed. In a day or two the eyes assume a reddish color, and the jaws, or skin on the sides of the jaws, become much swollen. When opened with a knife yellow water drops out freely. The animal commences rubbing the sides of its jaws against a tree, or anything it can get access to, and will continue to rub until both sides of the head are raw. In two or three days death ensues. Before it dies it becomes to all appearances perfectly mad and furious, but continues to rub its jaws until death relieves it of its sufferings.

Mr. H. CONLY, Cheyenne, Wyo., says:

We have some losses of horses, caused by their eating "poison-weed," properly known as "larkspear," which makes its appearance about the middle of April, and on or immediately after the 1st of May, in advance of the grasses. A hungry or jaded animal will eat a quantity of it, and within a few hours will begin to bloat, and if exercised or excited will tremble violently and fall down. If immediate relief is not obtained the attack will prove fatal. The remedies commonly used are lard and bleeding by slitting the animal's ears. Many of them will have all these symptoms and recover without any assistance. I am not prepared to give you the approximate losses caused by this weed. It is not so great as formerly, owing to the fact that stock-men guard against grazing upon lands where it abounds. I have taken some pains to test its properties, but owing to limited means have only found it to be acid, increasing in acidity as it grows older. I shall be pleased to send some of this weed at the proper time, should you desire it. If you could obtain an analysis of it, a simple antidote would follow, thereby saving us great losses.

Poultry seems to suffer about the same as in the States with diseases such as cholera and roup. All of the minor complaints come under the head of roup. We lose about 50 per cent. from the former. I believe there has been no sure remedy discovered. I have found that coal-oil will check the spread of the disease as speedily as anything, but that will not cure a fowl when attacked. I give it to them twice a week by saturating their corn and feeding them abundantly.

Messrs. CRAWFORD, THOMPSON & CO., cattle-dealers, Evanston, Wyo., say:

We have no general sickness of any kind among cattle, sheep, horses, or hogs. This is strictly a healthy country. The only disease among any of the above-named animals which has come under our observation is scab on sheep, which we very easily manage by washing with sulphur and tobacco.

Mr. FREEMAN WALKER, North Brookfield, Mass., says:

On some of our farms a large percentage of the cows abort their calves from one to four months before the time of calving. In some towns this has taken place from year to year, and the farmers know of no cause or remedy. I am expecting to come in contact with some of our farmers who have suffered in this way, and if I can get any facts of importance, I will communicate them to you.

Mr. T. N. BRAXTON, Paoli, Ind., says:

In this locality all classes of farm-animals, except hogs, have thus far escaped all diseases. The hogs in Southern Indiana have been diseased for the last year with what is known as cholera. Sometimes they are constipated, and at other times affected with vomiting and purging. Different remedies have been tried, but none of them have been attended with success. I have been feeding three hundred head this summer, in a lot in which there is a spring of running water. The water is very brackish, and leaves a sediment in the branch that looks like copperas. I also give my hogs ashes and salt. I have not lost one out of my entire lot with this disease. Other farmers in my neighborhood have had the disease among their hogs and have lost a great many. Salt and ashes may be a preventive, or the water from the spring may be; or, possibly, good luck alone may have caused my hogs to escape.

Mr. JOSEPH HOLE, Butlerville, Ind., says:

I am pleased that you are taking steps toward having an investigation of the causes of the fearful maladies to which farm-animals are subject. I shall be very glad to render you any assistance in my power for the furtherance of so laudable an object. Horses, cattle, and sheep are comparatively healthy. Chicken-cholera prevails to some extent, but not sufficiently to affect the interest.

The hog disease has prevailed in this and adjoining counties for several years, more, I think, as an epidemic than as a contagious disease. Cholera is a misnomer, so far, at least, as a large majority of the cases coming under my observation are concerned. Perhaps erysipelas or diphtheria would better describe the disease, although neither of these would, in all cases, be correct. What renders a description of the disease more difficult is, that while there are some general symptoms, such as loss of appetite with strong febrile tendencies, yet in a herd of hogs there will be a great variety of forms of attack. Thus several cases of sudden and complete paralysis have occurred. While the hog, previously in good health, was running for the feed prepared for it, it has been stricken down precisely as when shot in the brain with a bullet. In such cases, post-mortem examinations fail to discover any unnatural appearance of the intestines; but the condition of the lungs generally indicates strong symptoms of congestion.

One of the common symptoms that precede an attack of the disease is a dry, hacking cough. This, however, may continue for months without any further manifestation of the disease, though generally it is followed by the next symptom, a loss of appetite. And here any of the forms incident to the disease, and which no one can foretell, may set in. Intestinal fever is a common attendant of the malady. In my own observation the bowels, as a rule, are constipated, the animal passing only small, hard pellets. Very rarely fetid diarrhea is observed. About 70 or 80 per cent. of the cases prove fatal. Of those that recover, complete convalescence is not established under six or eight weeks, and even then no one will buy them if those that have never been affected can be had.

No panacea has been found for this terrible disease, nor has any treatment been tried, so far as I know, that can be recommended. The use of antiseptics is perhaps the best treatment. Bisulphate of soda, sulphate of iron, turpentine, charcoal, opium, nitrate of silver, carbolic acid, and creosote, have all been tried, and all have failed in bad cases. Preventives are better than cures. As a rule, those who give their hogs the best feed while in health, and look most carefully to their sanitary condition, escape with less loss than those who are less careful.

D. W. VOYLES, M. D., New Albany, Ind., says:

I am in receipt of your letter of the 10th instant, asking my assistance and co-operation in furnishing certain information in regard to the diseases coming within the range of my observation that affect the domestic or farm animals in this section of the country, and in reply would state that I fully appreciate the necessity of such a movement, looking, as it does, to the protection of the interests of such a large and important class of our population. The assistance I may be able to render you will prove of but little practical value at present, but may serve as an argument for the necessity of your undertaking, if argument is needed.

There has been no prevailing epidemic in this section of the State within the past few years affecting any class of farm-animals, except swine and fowls; and in these cases almost all complaints resulting in death are summed up by the people as "hog" or "chicken" cholera. In few cases of disease of this kind that have come under my immediate observation are found any symptoms analogous to the disease of that name as affecting the human species. The treatment has been as empirical and irrational as the diagnosis has been erroneous.

As to swine, the diseases, whether one or many, have created or caused a fearful loss to our farmers, and discouraged them in the further pursuit of that branch of stock-raising; and while almost all sections are more or less affected, there seems to be more disease and greater fatality among the farmers operating in the rich valleys of White and Wabash Rivers; and the disease and loss has been, I think, greatest during those years of great overflow, and greatest during the years in which the overflow occurred late in the season, leaving with its sediment the luxuriant growth of vegetation to decay and evolve miasmatic effluvia. From these facts I think much of the loss is caused from diseases brought on by exposure to miasmatic influence.

I am not conversant with any scientific investigations that throw much light upon the cause of these diseases, or the pathological conditions found on *post-mortem* examinations in such animals as have died. No treatment has yet been discovered by our farmers so certain in its curative effect as to inspire them with the belief that hog-raising is sufficiently safe from loss to insure profitable results. I am fully persuaded that so long as they content themselves in ascribing to all deaths the one common cause, "hog-cholera," and adhere to the present plan of empirical treatment, instead of patiently and scientifically investigating the causes producing disease in swine, and the various kinds of disease to which that animal is liable, giving to each its distinctive rational treatment, the subject will remain a mystery, and the fearful mortality continue to increase. That the farming population of the country, as a class, are not sufficiently educated to undertake this work, is a fact too well known to be disputed; and inasmuch as the great loss from that class of animals alone is not merely an individual loss, or the loss of a particular class of our people, but through them a great national loss, it is unquestionably within the range of the duties of the general government to undertake the extensive investigations which alone can accomplish this result.

What I have said in a general way in regard to swine applies with equal force to fowls. The loss, from whatever cause, is due to "cholera," in the opinion of most farmers, and astringent drinks and iron mixtures are given, whether the fowls are purging or are constipated from congestion from overfeeding, dying from starvation, or eaten up by vermin, or diseased from the foul air that arises from the filthy excretions remaining in their pens, unmoved for months, or from any of the many other causes affecting their health.

No fatal disease has prevailed epidemically among horses in this part of the country within the last few years, and this animal, therefore, is admitted by common consent to be liable to quite a number of ailments, requiring different causes for their production and slight modification in the administration of remedies for their cure; but in the case of the horse, the naming of a prevailing epidemic a few years since has unfortunately caused all bronchial and catarrhal affections to be grouped under one common class and name—"epizootic." In the treatment of this animal for whatever disease, we generally witness the heroic empiricism practiced upon the iron constitutions of the people of two centuries ago (who sometimes triumphed over both the disease and the doctor), by a selection of remedies from among the most poisonous and potential that can be found described in *materia medica.* All structural enlargements that do not warrant their removal by the surgeon's knife, instead of being slightly stimulated locally, in addition to such internal treatment as is calculated to favor their absorption and natural removal, are plied with blisters and the cautery until the country is filled with valuable animals scarred and crippled for life—living monuments of the ignorance and savagery of their owners and masters.

These facts, which I think are not overdrawn, show the impossibility of my giving you any tabular statement of diseases properly classified, and the treatment given under any proper head, because the several diseases affecting each class of animals have not been investigated, and are neither understood nor rationally treated. What I have been able to contribute, therefore, can serve only to show the great necessity for scientific investigation. The treatment of domestic animals in the West is generally committed to self-styled veterinary surgeons, whose experience is alone their guide, and that often founded upon the service of keeping some gentleman's horse, observation in a livery-stable, or as a common loafer around the neighborhood of breeding establishments.

Dr. H. J. DETMERS, vice-president of the American Berkshire Association, Bellegarde, Kans., writes as follows:

A thorough investigation of the epizootic, enzootic, and contagious diseases of live-stock is certainly a step in the right direction, and will be, if judiciously conducted, of more benefit to the farmers and stock-raisers of our country, and add more to the national wealth, than anything that has ever been undertaken or accomplished by the Department of Agriculture during its existence. I take the liberty of expressing my opinion thus freely, because of my almost constant residence in country towns and among farmers, my past connection with three different agricultural colleges in this country, and with one in Europe as professor of veterinary science, and still more, my extensive correspondence on the subject of live-stock and its diseases, with a great many farmers, and stock-men in nearly every State in the Union, who ask my advice as the conductor of the veterinary department of the Chicago Weekly Tribune, have given me ample opportunities to study the real wants and needs of the American stock-raisers.

With your permission I will remark that a rational and successful treatment, and still more a prevention of a disease, will be comparatively easy if the diagnosis is correct, and if the nature and the causes of it are understood and well known. If they are not, and if the diagnosis is doubtful, a successful prevention is impossible, and a rational treatment is out of the question. The best and most scientific prescriptions can do no good in the hands of any one who does not know what to do with them, and still the average American farmer craves for prescriptions like a child for sweetmeats, because he has the erroneous idea, fostered by quacks and charlatans, that experimentation will reveal a specific remedy for every disease, and that medical science has nothing to do but to label each disease and to search for a specific. Even learned medical men have made grave mistakes, and have delayed the progress of science considerably by their great confidence in drugs and their search after specific remedies which do not exist. In all diseases, but especially in epizootic, enzootic, and contagious diseases, a removal or a destruction of the causes, constitutes the only rational and effective prevention. Therefore, if the nature and the causes of those diseases are thoroughly investigated, and the result is laid before the public in language comprehensible to any man with the usual amount of intellect, a great many millions of dollars may be saved every year.

When our agricultural colleges were called into existence and most liberally endowed by act of Congress, I hoped and expected that a chair for veterinary science would be created in every one of them. Instead of that, men who have no interest whatever in agriculture, though some of them may be learned enough, become the presidents in nearly every institution. The machinery became very complicated; a great many things were taught which are of no use, either to the farmer or the higher mechanic; expenses accumulated, and no money remained available to teach those sciences most intimately related to agriculture. * * * * *

The investigation which you propose would have been made long ago, and many millions of dollars would have been saved to our farmers, if the agricultural colleges had complied with the law of Congress which commands them to teach those sciences related to agriculture, &c., or if a veterinary school worthy of the name were in existence in our country.

Mr. ELISHA GRIDLEY, Half-Day, Lake County, Illinois, says:

Stock here is generally healthy. Sheep were seriously affected with foot-rot a few years since. The diseased flocks were either cured or removed. A strong solution of pulverized blue vitriol, applied after thoroughly paring the feet, is one of the remedies used.

Cholera has not affected the hogs in this locality, but has destroyed large numbers of common fowls, turkeys, &c. Asafetida-gum, inclosed in a mosquito-bar and placed in the water they drink, has been used as a preventive, and, I believe, with favorable results.

Mr. JOHN E. THOMAS, Sheboygan Falls, Wis., says:

Farm-animals in this region are free from contagious and other prevalent diseases to which they are subject. Since the prevalence of "epizootic," some years since, our farm-stock have measurably escaped diseases of every description. Foot-rot in sheep, cholera in hogs, pip, &c., in fowls, and contagious diseases among horses, have all given us the go-by in later years. "Epizootic" is so well known that I do not regard a diagnosis necessary.

Mr. T. BACON, Wauconda, Ill., writes as follows:

Farm-animals in this county, so far as my knowledge extends, have been blessed with an almost entire absence of all diseases, excepting cholera among chickens. Poultry-yards have suffered terribly. All the nostrums have been tried, but with very poor success. Probably three-fourths of all the flocks infected have died. Even

Guinea fowls, roosting in trees, far away from all other fowls, have suffered in like ratio.

I have sometimes thought that hog-cholera was partially of a local nature, as we do not suffer to any great extent from it in Northern Illinois and in Wisconsin. If the ringer was abolished, and the hog allowed to use his natural propensity, I have no doubt the disease would be greatly abated.

Mr. WALLER BRODIE, Whitaker's, N. C., writes as follows:

The only disease existing among any class of farm-animals in this section at present is hog-cholera. We have been troubled a great deal this year with the disease. Over 50 per cent. of our hogs have died, and a great many are now affected. The first symptoms of the disease are manifested in the way of a fever, followed by a diarrhea, which continues about three weeks, when death ensues. But few infected hogs recover. The disease is more fatal with young than with grown hogs. A good many remedies have been tried, but with indifferent success. In my hands sulphur has proved more beneficial than anything else.

Our fowls have also been affected with cholera for five or six years past. This disease is also very fatal, and plays sad havoc in the poultry-yards. Calomel has been used, I learn, with some success. The disease is not so general as it is among the hogs.

Mr. JACOB GROVES, Boston, Mass., says:

The first symptoms of roup are those of severe catarrh or cold, followed by a peculiar and offensive discharge from the nostrils. Froth appears in the under corners of the eye; the lids of the eye swell, and in severe cases the eyeball is entirely concealed. The nostrils become closed by the discharges, which appear to be about the same as the excrements of the fowl when suffering with diarrhea. This last symptom is so well known that a description is deemed unnecessary. The cause of roup is a too scanty supply of grain, which necessitates an excess of green food.

Mr. J. EARL LEWIS, Pendleton, S. C., writing under date of September 16, gives the following case of bloody murrain in a valuable young bull owned by him, and his manner and successful treatment of the same:

His urine was of a dark, muddy color, and it seemed to give him great pain to urinate, which he did in very small quantities. After eating he would become very sick, and was not disposed to move about much. When he did so it would apparently give him much pain, and he would travel but a short distance, when he would again lie down. I gave him, in the evening, one pound of Epsom salts, three tablespoonfuls of saltpeter dissolved in flaxseed tea, using for the same about one-half pint of flaxseed. In the morning I gave the same drench, except that I reduced the Epsom salts to one-half pound. There was not much change in his condition, save in the appearance of his eyes and a freer discharge of urine. It was a week before any decided change was observable. I continued to give him daily a little flaxseed tea and saltpeter, which seemed to bring back the natural color of the urine. At the end of a week he recovered his appetite and would eat heartily, but his food seemed to make him sick, and he would vomit like a human being. I then gave him, once a day for two days, red-pepper tea and salt, which had the desired effect. He gradually recovered, but has never been in as good condition as before the attack.

Mr. THOMAS RUDD, Waukegan, Wis., says:

We have a disease here among cows called milk-fever, which has proved very fatal. The loss has been greater than from any other disease. "Bloat" and "dry murrain" also exist to some extent, but these diseases, I think, are attributable to irregular feeding. Many other diseases prevail here from time to time among farm-animals, a description of which I will endeavor to send you soon.

Mr. F. M. CORYELL, Brewersville, Ind., says:

Cholera exists among both hogs and fowls in this section of the State, and has proved very fatal among both classes of animals, probably nine cases out of every ten having proved fatal. In the first stages of the disease, as it affects swine, the symptoms widely differ. In most cases a loss of appetite is first noticed. Sometimes the animal is constipated, while in other cases exactly the reverse may exist. No cure has been found, and the only preventive of any value is to separate the sick from the well hogs at once, and give those not affected an abundance of salt.

The disease among fowls is equally fatal, the loss being about nine-tenths of those infected. Like symptoms prevail as in hogs, loss of appetite, followed by diarrhœa, which continues two or three days, when the fowl dies. No remedy has been discovered.

Mr. Z. T. MILLER, Raywick, Ky., says:

Hog-cholera is the most destructive disease we have to contend with here in Kentucky. It will attack a lot of, say, one hundred head of hogs, and in two or three weeks it will not leave a victim to prey upon. The disease is more general and much more fatal in some localities than in others. A gentleman living in Nelson, an adjoining county to this one (Marion), has not had a case in a lot of two hundred head of hogs, while his neighbors have lost from three hundred and fifty to four hundred head. Why should this be the case? Perhaps feeding has something to do with it. Upon inquiry, I learn that the gentleman whose hogs have escaped the disease feeds cooked meal, in which is mixed wood ashes, char and stone coal, sulphur, copperas, and coal-oil. This has been a successful preventive. His neighbors, who feed nothing but dry corn, have suffered severely. Dry corn is too stimulating, and produces fever. This is soon followed by loss of appetite, and the next symptoms are those of cholera. It is then too late to commence drugging them, as they are almost sure to die. However, a few might be saved if the sick were separated from the well ones as soon as the first symptoms of the disease were discovered. If hogs liable to infection were fed on cooked meal with the mixtures above named, I am inclined to think they would escape the disease.

A peculiar disease has recently broken out among the horses in this locality. Its first symptoms are observed in a severe stiffness of the joints of the animal, so much so, indeed, as to render him unfit for service. In a week or ten days his body will become greatly swollen, and he will break out in sores from which an offensive matter will be discharged. The disease does not seem to prove of a fatal character, but the horse is seldom worth anything after a severe attack.

Hon. DANIEL M. HENRY, Cambridge, Md., says:

From time to time I have heard of the diseases commonly known as "blind-staggers" in horses, and "hog-cholera" and "chicken-cholera," prevailing in localities of greater or less extent with great fatality, but they do not seem as yet to have produced a professional veterinary surgeon.

Mr. THOMAS STURGIS, of the firm of Sturgis & Goodell, cattle-breeders, Cheyenne, Wyoming, says:

Among cattle but one disorder is recognized by stockmen here. It is known as "blackleg." It occurs but seldom, and is confined solely to large cattle and yearlings. Its attacks are most frequent among fat cattle. It kills in twenty-four to thirty-six hours. The symptoms are stiffness of shoulders and chest, and swelling of legs from above downward. Recoveries are few, if any. No treatment has been found successful. The largest loss known was that sustained by Edward Creighton, of this Territory, who, some years ago, lost two hundred head of calves and yearlings.

Spanish or Texas fever is unknown here. The probable cause of escape is owing to the state of nature in which cattle live—absence of prepared food or shelter and extensive ranges, where they are widely scattered.

Horses have suffered greatly from an epidemic closely resembling the epizootic distemper experienced in the States two years ago. They grow very weak and thin, cough, and discharge at the nostrils. Some die. In most cases recovery takes place in three or four months.

Scab in sheep is known here, but the disease is not widespread. The remedy employed seems effectual, viz., a solution of tobacco used as a dip, and repeated until a cure is effected. A dip made of a solution of carbolic acid and water is also employed, as are other dips in which arsenic mingles.

Any remedies known to be valuable for diseases of poultry will be gladly received. The symptoms of commonest trouble are dull eyes, unwillingness to move, failure to eat, and death in about three days.

Mr. DANIEL CHAPLIN, La Grande, Oreg., says:

There is no prevailing disease among domestic animals in this county except that of scab in sheep. This parasitic disease has heretofore prevailed to considerable extent on the Pacific coast, but it is getting to be better understood and is now fast disappearing under close watchfulness and timely remedies. I have had considerable experience with the disease, and have succeeded in exterminating it as follows:

First, I sheared my sheep very close, and peeled off all scabs or sores, and at the time of shearing I spotted every diseased place with a strong wash of corrosive sublimate and water. I then dipped them three times in a strong decoction of tobacco, using one-half pound of stem-tobacco to each sheep. The dip was heated to 120° F., and the sheep held in it at least two minutes. The dipping should be performed at intervals of fourteen days.

After several years of experimenting, I found this a sure remedy. Many other prescriptions were used, but without success. The sheep should be put on fresh pasture after dipping, and not allowed to run on the old one for one year thereafter. Scab is the only disease to which sheep are subject on this coast. Foot-rot and other diseases so prevalent in other localities are entirely unknown in Oregon. There is no prevailing diseases among other animals.

Mr. J. N. SMALLS, Scotland Neck, N. C., says:

We have but very few diseases among farm-animals in this section of the State, with the exception, perhaps, of the disease known as cholera among hogs. This disease assumes different forms or symptoms. Some are attacked with vomiting, and linger about one week; others lose their appetite, become sleepy, and their eyes become inflamed and exude an offensive matter. Cases of the latter class have been known to die within six hours after the attack was first observed. Many remedies have been tried, among others salt and ashes, tar, saltpeter, and bluestone. While some of these remedies have proved of value on one farm, on an adjoining one they have been found of no service whatever. The disease is not so prevalent as in former years, though it is, perhaps, more fatal. One farmer has lost sixty-three head out of a lot of sixty-five.

The disease among fowls is also called cholera. The first symptoms are drooping, loss of appetite, &c. They die within a few days. A lady friend has used calomel and opium with success. She administers it in small doses three times a day.

Mr. DONALD MURCHISON, Toulon, Stark County, Illinois, says:

For several years past the hog-cholera has annually been destroying immense numbers of hogs throughout various sections of this country. The probability is that $250,000 would not cover the losses in this county alone, since the disease first made its appearance. I am a farmer, and have been extensively engaged in hog-raising, and therefore have given the disease a great deal of careful thought and study, and I believe I have now found a sure remedy, if given in the first stages of the disease. The recipe is as follows:

Make a strong tea of smart-weed. After the weeds are separated from the solution, add one-eighth of a pound of arsenic and one-fourth of a pound of concentrated lye, and from one-fourth to one-half spoonful of flaxseed to the hog (according to the size of the animal). It is best to have the solution boiled over again after the flaxseed is added, or else have the flaxseed cooked in a smaller vessel and thoroughly mixed with the tea after the weeds are separated from it. Then mix a sufficient quantity of oats in the tea to soak it all up, and feed to the infected hogs night and morning as much of it as they will eat. Give them no other food for a week or ten days, or until they begin to show unmistakable signs of returning health, when this feed may be gradually lessened and corn given in its place, gradually at first, and increased as the other is diminished.

The flaxseed is not necessary unless the bowels are constipated, which is generally the case in advanced stages of the disease. Some farmers use salts when the bowels become constipated; but it is about the worst thing that can be given, as it is a blood-cooler and a blood-weakener. Although it may give temporary relief it prostrates the system, and in a few days the hog will be in a much worse condition than it was at first. The flaxseed is a good laxative, and at the same time is very nourishing and strengthening to the system. The lye seems to be a good tonic as well as a good remedy for cough in hogs. The arsenic also acts as a tonic; besides it kills the worms with which all hogs (with very few exceptions) are troubled. The smart-weed is probably about the best remedy for inflammation that we have. It is warming and strengthening to the system, and gives tone to and equalizes the circulation of the blood, just what seems to be needed, as the disease is a congestion of the lungs.

Mr. A. B. McKEE, Vincennes, Knox County, Indiana, says:

There are but two diseases, so far as I know, that prevail as epidemics in this section of country, viz., hog and chicken cholera. The hog-cholera has become one of the most serious diseases with which the farmer has to contend. He may think he has a fine killing for the winter, but the cholera enters and in a few weeks he finds himself left without enough for his own family supply. The disease presents so many different phases as to prevent me, with the little investigation I have given it, from attempting a complete diagnosis. A drooping of the head, loss of appetite, and an indisposition to move are among the first symptoms noticed. Sometimes there is a running off at the bowels, and sometimes constipation prevails; sometimes they die in a few days, and then again they may linger for weeks. I confess I do not understand either the pathology or the workings of the disease. As to the cures recommended, they are numerous, and generally based not upon a scientific analysis of the remedies prescribed but upon the vague conceit of the party recommending them; and then, again, all the different remedies in turn have proved failures. If Congress would do anything to throw light on this subject, and especially if a specific could be found, it would prove of incalculable benefit to the whole country.

I have used, and I have thought with some benefit, alum and Venetian red—alum as an astringent and venetian red as an absorbent. During the past summer, I have used poke-root, given in slop, in such doses as to secure its alterative effects, and as a preventive rather than a cure. From its known effect as a preventive in other diseases I have no doubt as to its beneficial effects in this.

Mr. W. W. WOODYARD, Morristown, Shelby County, Indiana, says:

We have no disease affecting farm-animals in our locality, except cholera among hogs, or a disease making its appearance in many different forms, called cholera. In some cases the pig, when quite young, will become affected about the eyes, and partial or total blindness will follow in a few days. A high state of inflammation about the mouth and throat next makes its appearance. Perhaps 90 per cent. of such cases will prove fatal in from six to ten days.

In other cases the hog, at a greater age, will first show signs of inflammation about the ears and neck, the ears becoming sore, with a yellowish mucus making its appearance about the root of the ear. Very few of those attacked in this manner recover. Some will simply show a sleepy, sluggish appearance, refuse to eat, and usually die in from twelve to twenty-four hours.

Many remedies have been used, but the best informed men will say, almost unanimously, without the least benefit whatever. The disease is prevailing to a fearful extent in some localities at this time. The president of the First National Bank at Rushville, Ind., who has large opportunities of knowing, says that Rush County alone will lose $500,000 by this disease the present year. Our own county will perhaps be equally as heavy a loser.

Mr. JAMES FERGUSON, Ashborough, Clay County, Indiana, says:

For fifteen years, at intervals, what is known as hog-cholera has been very destructive among this class of farm-animals here. Personally, I have had but little experience with it. In some the symptoms are refusal of food, stupor, apparently nearly deaf and blind, constipation, and death within from one to five days. Others have vomiting and lax evacuations, of which seven-tenths die soon.

Of the cause of the disease I know nothing certain, nor have I heard a rational theory from our farmers. Various drugs are administered as long as the hog survives the disease and the doctoring. I know of no reliable remedy.

Apparently it is safe to assume that worms, and, possibly, other parasites on the digestive organs are the cause of most hog diseases. Hogs that have frequent doses of sulphur, copperas, turpentine, and arsenic, with free access to wood-ashes and charcoal, are usually healthy, and almost exempt from cholera.

Chicken-cholera is not unknown among us, but I think its cause and remedy are alike unknown in this locality.

Mr. LEWIS J. REYMAN, Salem, Washington County, Indiana, says:

We have no prevailing disease among farm-stock in this county, except hog and chicken cholera, which has prevailed for quite a number of years, and is prevailing to some extent at this time. Two years ago this fall I turned thirty-five hogs of my own raising in a corn-field, and they fattened very fast for about three weeks. About this time I bought twelve head that were raised on low, wet, river-bottom land, about twelve miles from my own land. In a few days some of them were attacked with cholera, and two died. In a few more days those of my own raising took the disease, and nine of them died. The balance lost flesh for a time, but gradually recovered. I fed them sulphur and ashes, calomel, May-apple root, and a number of other remedies recommended.

There are various symptoms of the disease. In some instances they vomit and purge, and in others their lungs seem to be affected, and they are constipated. When the lungs and bowels are affected they seldom if ever recover. Those that are attacked with vomiting and purging get well, and soon go to fattening again.

The same fall (two years ago) I had thirty pigs that were just weaned. They were taken with purging, and all died. I also had about thirty spring shoats that were affected in the lungs and bowels. They all died but four or five. Two lingered for some days and would not eat grain, but would drink a little milk. I concluded I would experiment a little with these, and I gave them each one tablespoonful of sulphur every morning in their milk for two weeks, when they commenced eating corn again, and gradually got well.

Mrs. L. J. REYMAN, of the same county, says:

Chicken-cholera has prevailed in this neighborhood for years, in a majority of cases proving fatal to the whole flock. Three years ago our fowls had it for the first time, and out of about forty hens and several dozen young chickens (that were hatched late in the season), I had but ten hens and a few chicks left. I tried several remedies, feeding them asafetida, Cayenne pepper, alum, &c., but do not know that they had any effect, as the chickens mostly dropped dead off the roosts in the night, being apparently well the day before.

About two months ago my flock were again attacked with the disease. Only six or eight died. A few of them dropped off the roost dead, and some lingered near two weeks, eating a little, but getting weaker until they died. A few recovered after being sick several days. I used nothing but a little cayenne in their feed and alum-water by them constantly, mixing their feed with it also. About the same time a tenant on another part of the farm lost about forty large fat hens and some turkeys. She saved about one-fourth of her flock, including the young chicks. I do not know what remedies she used, but I do not think she used any alum. This disease, for the last few years, has had a depressing effect on the market of poultry and eggs in this county, and we are needing a remedy badly.

Mr. DUNEHEW, Trinity Springs, Martin County, Indiana, says:

There seems to be no serious disease among any class of farm animals in the county now, excepting among swine. Mr. F. F. Sholts, an experienced farmer and stock-dealer, says in regard to lung diseases of hogs, that it begins with a cough, which increases for two or three weeks, and if not arrested by that time the lungs will be so decayed that death will ensue. The cough generally begins by a kind of snuffing, as if dust had been inhaled. At this stage, a few doses of his medicine will cure every case. The longer it runs the more difficult it is to cure. "The old-fashioned cholera," he says, "puking and purging," is also readily cured by him, and if promptly handled' the cure is speedy and effective. He claims that if the medicine is given as a preventive no case will occur. Mr. Sholts will not make public his remedies, as he has persons now traveling engaged in selling his medicines.

Mr. W. DANIELS, Huntington, Huntington County, Indiana, says:

There is no disease among farm stock here, except cholera in hogs, and this is not so prevalent as in former years. No remedy has been found.

Mr. JOHN M. BARNETT, Somerset, Pulaski County, Kentucky, says:

Aside from hog cholera, our farm animals have been free from disease, except the usual amount of cholic and grub in horses. Our cattle have been entirely free from disease. My experience, as well as that of others in this section, is, that pine-tar given to hogs will act as a preventive in localities where hog cholera is prevalent.

Mr. J. B. MILLER, Hartford, Ohio County, Indiana, says:

We have in Southeastern Indiana a disease among hogs and chickens called cholera. It is sometimes very destructive among hogs. I have found the disease, by dissection, to be confined principally to the lungs. This some produces a dry, scorching fever, which thickens the blood and causes death. The disease is caused by worms and an accumulation of dust in the lungs. It can be prevented by putting soap-suds in the slop-barrel and feeding them a quantity of it once or twice a week. Soap is a good remedy for worms, and also cleanses the bowels. I have cured several in the first stages of the disease with turpentine and coal-oil, using it in equal parts and giving three drachms once a day. I also rub it on the center of the back behind the shoulders once a day. If these remedies are properly used the disease need not be feared.

Hogs in the Western States do not have proper attention. Too many are allowed to sleep together, when they get overheated and die of thumps, which is also called cholera. If people would take better care of this stock, and use soap in slops as above recommended, hog cholera would soon pass away.

Chicken cholera is a disease of the liver. Only one cure is now in use in this county, which, if taken in time, is a specific. Take May-apple root and boil a weak solution. Add a teacupful to a quart of meal and feed. Corn lime is a good preventive

Mr. J. HARBISON, Charlestown, Clarke County, Missouri, says:

I would have answered your letter sooner, but I have been waiting to see some of my neighbors, one of whom has lost eight head of stock cattle in the last three weeks; but he could not tell me much about them, only that they were covered with ticks, that they would not eat or drink, and that they would stand for hours at a time with their heads to the ground, in which position they would remain until they died. The cattle were bought at the Louisville (Ky.) stock-yards by Mr. H. J. Crum.

I have lost some fowls from roup and cholera. Of the two diseases I dread roup the most, as it does not show as unmistakably as the cholera. A fowl with the roup will eat heartily, and to all appearances look well, until the disease will break out among the entire flock. They hardly ever die with it, but they lose their eyes and look so disgusting, that I generally cut their heads off as soon as I find them affected with it. In fact, that is my remedy for all diseases, especially cholera. Cholera usually shows itself by the fowl moping around, generally with a full crop, sometimes with nothing in its craw; will not eat, but drink often; the comb and wattles become a dark red—nearly a black color; the discharges at first are a pale green color, then dark green, and sometimes yellow, like the yolk of an uncooked egg. They are generally fat when taken, and seem to die sooner than when in a lean condition, I have sometimes found the fowls with their craws so full of dry grass that it would not pass beyond. By cutting open the craw and taking out this food, washing with warm water and sewing it up again, they will soon get over the disease and in a few days will begin to eat heartily.

Mr. JAMES A. LEE, Dowagiac, Cass County, Michigan, says:

Stock in Michigan is subject to but few diseases. Horses generally have the distemper when growing, which runs from one to two weeks. It commences with a slight cough, watering of the eyes, and loss of appetite. As the disease progresses the cough increases. The throat and jaws swell, gather and break, when the horse becomes unable to swallow, and dies of suffocation. The disease will yield to mild treatment, such as sweating of the head and throat with bitter herbs, viz: wormwood, catnip, hops, &c., and smoking the head with sulphur and old shoe-leather. Take a ball of good sweet butter as large as an egg and put it down the horse's throat twice a day. Give mild physic if the case needs it, and keep the horse warm.

The most common complaint among horses is cholic. The symptoms are extreme uneasiness. The horse paws, lies down and rolls, gets up and lies down again, groans, begins to bloat, and continues in this way until death ensues unless relieved. This can invariably be effected, if taken in time, by a dose of one-half pint of salt dissolved in a quart of water. It should be administered every ten minutes until relief is afforded, which generally occurs after the third dose.

Worms of different kinds affect the horse and are very troublesome. The symptoms are tight skin, rough coat, irregular appetite, and the appearance of a yellowish mucus under the tail. The horse lifts one hind foot to the belly, draws himself up, partly lies down—perhaps entirely down—on his belly, gets up immediately and goes to eating, stops suddenly, and does the same thing over again. Give a common tablespoonful of copperas in a ball, followed by a bran-mash once a day until relieved; then give a light physic or turn to grass.

We have nine head of cattle affected with horn distemper to one affected with any other disease. I am well aware that there are many learned men who say there is no such disease, yet I know by forty years' experience the truth of what I write. The animal has a staring look and a yellowish deposit in the corners of the eyes next the nose, grinds its teeth, hair stands up, tail soft two or three inches from the end, bowels varying from costive to laxative. This continues sometimes for years, and is attended at times with loss of appetite and strength. The pith of the horn decays and is discharged at the nose, and finally the membrane over the brain gives way and death ensues. Cure: Cut two inches off the end of the tail to start the blood, and the bone will be found lacking. Take one-half pint of sharp cider vinegar, put in a tablespoonful of black peper and same of salt; dissolve well. Then take the animal by the horn and nose while some one injects one-half gill of the liquid in each ear. If very bad, so that the animal is down, bore the horns with a spike-gimlet, and inject some of the liquid with a syringe. Give a tablespoonful of copperas and saltpeter in a ball or mash every day for a week, then every other day for another week. Sometimes a cow will be in full flesh and drop a calf in midsummer, give plenty of milk, and do well for a few days; the next day give no milk, and perhaps not be able to get up at all. For this trouble give the above treatment, with an occasional slice of fat pork. Let the chill be taken from the water she drinks, and a cure will be effected. Garget is a very troublesome disease in milch cows. The cow becomes feverish, the udder especially. Sometimes the milk will be streaked with blood, and again appear lumpy, or both. Wash the bag with bitter herbs steeped in vinegar. Give a tablespoonful of poke-root, pounded fine, in a bran-mash, twice a day. Also, insert a seaton in the brisket with a piece of poke-root.

Hog cholera is increasing to an alarming extent. The first symptom is generally observed in the animal carrying his nose near the ground, with a generally dull appearance, slight cough at first, and swelling under the throat. Some are first taken with severe purging. All these symptoms increase in intensity until death ensues, which usually occurs in from twenty-four to thirty-six hours. No cure has yet been found. Strong wood-ashes and copperas are regarded as preventives. I have cured some by drenching with copperas, sulphur, and asafetida.

JAMES C. DUSTAN, V. S., Morristown, Morris County, New Jersey, says:

The appearance in this section of a new and unusually fatal disease among horses has prompted me to report to you some of the facts connected therewith. It may be more common in other parts of the country, but here it is new to our profession. The first case occurred about the middle of last month in the adjacent village of Madison, and up to the present time twenty-one horses have been attacked by the disease in that place. Of that number eight were under my professional care. Four of these have died and the others have recovered. Of the remaining thirteen, only one has recovered. The disease is of short duration, lasting, in the cases that prove fatal, from two or three days to one week. The general symptoms are as follows: For the first day or two the horse seems inclined to droop, and, without any apparent cause, acts tired. Then a difficulty in swallowing is noticeable, which increases as the disease advances. The fever is high; obstinate constipation of the bowels, and almost complete suppression of the urine, the latter fact being ascertained by means of the catheter. The manner in which the act of swallowing was effected made it clear to my mind that the inability to do so was caused by a partial paralysis of the muscles of deglutition. Generally, when the horse lies down, he is unable to rise without assistance. There is also a marked tremor in the left fore shoulder, and, although not constant, has been noticed by me in all the cases I have seen. Post-mortem examinations in four cases have disclosed the following anatomical lesions: The most prominent is an intense inflammation of the larynx, extending for some distance down the trachea. The kidneys were found to be in a state of congestion, and in one case considerably hypertrophied. There was also found inflammation in the nasal fossæ, but more particularly in the left. The brain, œsophagus, and spinal cord were found in a state of perfect health, as were also all the other organs of the body. I regard the disease as one of blood-poisoning, introduced into the system from the atmosphere, and, as far as I have been able to ascertain, it resembles in a striking degree diphtheria in the human being.

My treatment consists, first, of a blister of cantharides applied to the larynx region, and kept open for several days by mild mercurial ointment; dry cupping over the kidneys; the administration of linseed-oil as a laxative, aided, if necessary, by injections, and the following prescription given every four hours, viz: five drops extract of belladonna, one ounce of water, and one-half drachm of iodide of potassium. This is for one dose.

To the above prescription was added, for fever, tincture of aconite, and after a day or two, dropping the aconite, I gave quinine sulph. grs. x, every three hours. The use of iodide of potassium should be continued until the functions of the kidneys have been fully restored. I also found benefit from the free use of chlorate of potash. The diet should be of the most nourishing kind, and by every possible means the strength of the animal should be supported. As a drink, hay-tea is preferable to plain water. But the best treatment I could give, together with careful nursing, shows as a result a fatality of 50 per cent.

Hon. A. L. PRIDEMORE, Jonesville, Lee County, Virginia, says:

The hog cholera has been among us for years. It was first brought here by drovers from Kentucky. It prevails to a greater or less extent every year; some years killing all (or nearly all) the hogs in some neighborhoods. I have known instances where three hundred head of hogs were turned in upon corn-fields, the usual mode of feeding here, say August 20, and by the 1st of September the disease would break out and they would die at the rate of eight or ten daily, until, perhaps, there would not be twenty hogs left. Some of the diseased animals would pine away and dry up, as if with a fever; some would die with spasms; in others, the flesh would, so to speak, rot and slough off to the very bone; a foot, ear, tail, or nose drop off. Some would have symptoms like a person with violent vomiting and purging. These rarely ever recovered with us. This year in this county the disease, thus far, has been very mild.

Poultry are subject to a disease also called cholera here. They drop off the roost dead, and fall over suddenly in the day-time. Sometimes the disease will kill all the fowls on a farm at the rate of eight or ten per day. No remedy has ever been found for either of these diseases. Their causes are wholly unknown to us, though there are many conjectures.

Mr. A. C. ELLIS, Hartford, Van Buren County, Michigan, says:

Hogs have been affected with various diseases in this locality. The disease commonly known as cholera is, perhaps, the most fatal. Its duration is generally short, fatted hogs usually surviving but a few hours. The remedies have been various and generally unavailing, so far as my own information extends. As the result of frequent experiments I am prepared to give, as the most effective remedy I have ever seen tried, a simple receipt furnished me by one of the largest hog-dealers in this county. It consists of one pound of lime, one ounce of spirits of turpentine, and one-half ounce of pulverized saltpeter, mixed with soaked corn or slops. This amount is sufficient for twenty hogs, and, for a cure, should be given every other day; as a preventive, about twice a week.

There is also a fatal disease among hogs here similar to quinsy, affecting the throat principally. It is generally confined to young hogs. I know of no remedy. They usually survive from ten to fifteen days. At least 50 per cent. of the cases prove fatal.

No disease is common among sheep, save the old one, known as rot. Experience has proven that one of the best recipes for this trouble is the killing off or clearing out of the older ones of the flock, and frequent changing and crossing of stock.

Of the feathered tribe, chickens and turkeys are both subject to a disease known as cholera. The remedies used are various but unavailing. The duration of the disease is generally short; in fact, the attack is frequently instantly fatal. At least 90 per cent. of those attacked die.

Mr. S. WOODWARD, Ohio County, Kentucky, says:

There is but one class of animals in this locality subject to diseases which have not been common for years, and that is swine. A great many of these animals have died of a disease called cholera. The symptoms are various. Sometimes the hog will be attacked with purging, and again extreme constipation will prevail. Sometimes the hog will die suddenly, and in other cases it may linger until it becomes a skeleton. On an average about one-half of those attacked with the disease die. There are seldom any remedies tried, as we think they do no good. The disease is not confined to any certain season of the year. In this locality we have had none of it since January and February last, but we hear it mentioned as existing in other localities.

I am of the opinion that a great many hogs die from the effects of lice, especially those lingering cases. These vermin find them an easy prey in their weakened and emaciated condition.

Mr. G. H. LUCAS, State Line City, Warren County, Indiana, says:

A prevalent and fatal disease among hogs in this locality is known as cholera. The first symptoms are running off at the bowels, which is generally accompanied by a hacking cough. The animal becomes stupid, and refuses to eat. As the disease progresses it becomes very poor and emaciated, and stands around with its body drawn up as if in pain. The disease usually proves fatal in from one to four days. The following remedy for the disease has been used with moderate success in this neighborhood, viz:

One pound each of sulphur and madder, one-half pound of saltpeter, one-fourth pound of antimony, and one ounce of asafetida. This should be mixed with a pailful of slop or milk, and three tablespoonfuls given once a day.

I am satisfied the disease is contagious, and all infected hogs should be removed from the well ones, and those that die should either be burned or buried very deep in the ground.

Mr. H. SHUGART, Marion, Grant County, Indiana, says:

There is a very destructive disease among hogs here, called cholera, but in my opinion it is lung-fever. No remedy has been discovered that I am aware of. It is said that hogs do best, and are less liable to be attacked by the disease, that have clear, running water to drink, and are kept from a mud-wallow. This is a mistake, as more hogs die from the disease that are kept along water-courses than among those that are kept at a distance from creeks.

We also have what is called cholera among fowls. Many of those engaged in raising poultry are of the opinion that most of the diseases among fowls are brought on by the presence of lice. A thorough and frequent wash of the coops and sheds with strong lime-water will soon cause the disappearance of this vermin. The boards of the sheds should be set upright, and the edges not allowed to touch each other; then, if the roosting-poles are smoothly shaven and the above wash frequently used, the fowls will never be troubled with lice.

Mr. GEORGE W. ADAMS, Leavenworth, Crawford County, Indiana, says:

There is no prevailing disease among farm-animals in this county at present, except among hogs, in which class an occasional case of cholera is reported.

Mr. JOHN K. BEVIS, Taylorville, Bartholomew County, Indiana, says:

I will give you my own experience with the hog-cholera, as it is called. It first made its appearance on my farm in September, 1857, when I lost sixty head. I examined quite a number, and found them all spotted on the belly, and the throat full of clotted blood. It appeared again on my farm in June, 1875, when I lost ninety head. The disease worked different from its course in 1857. Some would lose their appetite and dwindle away to mere skeletons before death ensued, while others would die in a few hours; some would squeal as if in great pain, and would soon die; others, again, would take spasms, which would last for some days, and then die. I used copperas, sulphur, madder, turpentine, antimony, coal-oil, in fact all the remedies that I could hear of, but without effect; at least, all that I doctored died.

Recently, I have come to the conclusion that the rooter on the nose was put there for a purpose, and have not rung or cut the nose of any swine since. I have no reason to complain, as my hogs have since done well.

As to chickens, they seem to all die on the roost, as they are found dead in the morning. Since I commenced giving them copperas and sulphur in their feed and water I have had very good luck.

Mr. ELIZUR SMITH, Lee, Berkshire County, Massachusetts, says:

My cows are more or less subject to fouls in the feet in wet weather. I make use of tar and spirits of turpentine, but if you have a better remedy I would be glad to have it.

Mr. C. A. ADAMS, Chillicothe, Livingston County, Missouri, says:

This year the diseases among swine have proved more fatal among young pigs and fat hogs, farmers having lost from one to one hundred head. The remedies are so various and unreliable that they are not worth mentioning. It is very doubtful, indeed, if a remedy ever will be discovered until the sanitary condition of the hogs is improved. The animals are affected in different ways—some purge and vomit, while others cough violently. Some refuse to eat, and soon pine away and die. As the disease is most fatal in large herds and in filthy surroundings, it would seem to the interest of the farmers to look to the natural cause; and, if possible, remove it by confining a less number together and keeping their surroundings clean. Feed charcoal freely, and on the first symptoms of a cough give one teaspoonful of red pepper to each one hundred pounds weight of the animal. Given in slop, it has proved very successful with me.

I have lost but few hogs since I removed them from the old straw-stacks, manure-piles, old sheds, and from under old buildings, where their quarters could not be cleaned out. A change of feed from corn will always prove beneficial among hogs when diseases are prevalent.

Poultry will come under about the same conditions as hogs. Too many are kept together, and too little attention is paid to cleanliness.

There is no veterinary surgeon here who has given any particular attention to diseases incident to hogs and chickens.

Mr. U. F. GLICK, Columbus, Bartholomew County, Indiana, says:

Hog cholera, the prevailing disease among swine in this locality, is generally caused by impurity of air, foul feeding, filthy pens, &c. The disease is soon banished by clean pens, pure water, and cleanly habits.

Cholera is also prevalent and very fatal among fowls. This disease is also the result of foul pens and improper feed.

Mr. A. M. SANDERSON, Leesburg, Kosciusko County, Indiana, says:

A friend of mine had a fine lot of hogs this fall, varying from pigs to fat hogs. He has lost nearly all of them by some disease, probably the cholera. They were on wheat stubble after harvest, and then on clover pasture. When first taken their evacuations were dry and hard. This condition continued about three days, when diarrhea would set in, and they would die within a few hours thereafter. Nothing was found to do them any good.

There is a new disease in this locality among horses, called by farriers pink-eyed distemper. The horse, within a few hours after the attack, will become stone blind. Some get over it, while others only partially recover their sight. The eyes matter and run a great deal. The treatment thus far has been merely experimental—what would seem to relieve one would not benefit another.

Chickens in some localities have nearly all died of cholera. In my own experience I have found sulphur the only remedy. Mix with corn-meal and feed. With this remedy I have cured fowls that could not stand up.

Mr. O. W. HANNUM, Leavenworth, Crawford County, Indiana, says :

I have been dealing in stock in four or five different counties, and find all classes of farm animals healthy except hogs. A disease exists among this class of stock which carries them off very fast. Many people regard it as a malignant type of lung fever or pneumonia. I lost forty-five head of hogs by this disease one year ago. No remedy is known.

Mr. JAMES A. MARTIN, Salem, Washington County, Indiana, says:

I know of no disease among horses as fatal as the lung fever. I had one die with it some time since. It was taken with a hacking cough and difficult breathing, and lived seven days. I know of no remedy that I can recommend, for all die that are attacked by the disease here. This disease is not as common as the bots, but is more fatal. A lump of alum, the size of a walnut, given to a horse, will generally cure the bots.

The only disease existing among hogs is cholera, and there are various cures for it. Equal parts of sulphur and copperas, mixed in sweet milk, is the most effective remedy I have tried. Some have tried coal-oil, castor-oil, poke root, and also patent medicines. None of these remedies, however, are regarded as a sure and permanent cure.

Chicken cholera is common among the poultry here. Soot mixed with corn dough is the best remedy we have tried.

Mr. J. D. GUTHRIE, Shelbyville, Shelby County, Kentucky, says :

Hog cholera, in its incipient state, with shoats and half-grown hogs, usually begins with constipation, a symptom easily discoverable by their droppings, which are hard and marble-like. This is followed by a dry, hacking cough and internal fever, which increases as the disease progresses. These symptoms are attended with a gradual loss of appetite. At this stage of the disease their movements become listless ; they droop their noses toward the ground, and are shy of approach. The duration of these symptoms depends upon the severity of the attack. In some cases they continue six or eight days, and in others two or three weeks, with gradual loss of flesh until they look like walking skeletons. I refer now to the premonitory symptoms. After the disease becomes epidemic they frequently die within twenty-four hours after the first indications manifest themselves, without any regard to flesh or previous condition. In the latter stages of the disease they have a loose, discolored discharge, which soon terminates with thumps. This is a palpitation in the flank at the drawing of each breath. At this stage the disease is easily imparted to others, having become epidemic in form, and so fatal as to carry its victims off within a few hours. The remedies that prove efficacious in the first stages of the disease are worthless in the last. I would here recommend the removal of the diseased hogs from the rest of the herd, and the remedies hereafter mentioned given to the remainder. From my standpoint I am of the opinion that cholera in swine, in the last stages, is incurable, unless it be in isolated cases. I hold that constipation of the bowels is cholera in an incipient state, and whatever remedies would remove the cause the effect must necessarily follow. I speak only from my own experience and observation, which practice has fully demonstrated to be, in the main, correct.

I have been very successful in relieving my herd of constipation by giving one-half pound of calomel to fifty shoats, on corn moistened so that it would adhere to the grain. This should be repeated at intervals of twenty-four hours, until the bowels are opened by the medicine. Old bacon, grease, or linseed-oil will have the same effect, the only difference being that calomel will regulate the liver, while the others will only relieve constipation. Grease or linseed-oil, if given in doses of one-half pint, will cure thumps, which is the last stage of cholera in a constipated form. Hogs fed on apples, pumpkins, or following after cattle fed on corn, are not liable to cholera. A neighbor last spring purchased one hundred head of stock-hogs from the pens at Louisville, Ky. Soon after getting them home they commenced dying at the rate of four or five per day. He procured a large kettle and commenced cooking and feeding the dead to the living hogs. The result was that he saved sixty out of the lot of one hundred. Others have fed the dead carcasses of sheep, cattle, and horses to hogs affected with cholera, and the result was a cessation of the disease. Another acquaintance keeps his hogs well supplied with wood-ashes and salt, at the rate of two parts of ashes to one part of salt, which he says is a certain preventive. As all these remedies have the same tendency, namely, the opening of the bowels, we can consistently arrive at but one conclusion, and that is that the premises are correct and the applications act as an antidote to the disease known as hog cholera. I sincerely trust your inquiries may result in the discovery of something that will stay the further progress of a disease fraught with so much injury to the agricultural interests of our country.

The first symptoms of chicken cholera are observed in discharges of a thin, yellow nature, followed by an inclination to sleep, whether sitting or standing. These conditions continue until two or three days before death, and during this time the disease is very contagious. The most simple remedy is to give the flock water well impregnated with alum. This will usually stay the malady. The sick ones should be given a pill of pulverized alum the size of a small pea, inclosed in wheat dough. If the first does not produce the desired result, repeat the dose at intervals of a few hours. The fowl, when laboring under the disease, has a high internal fever and insatiable thirst, but no appetite or desire to eat.

Mr. JOHN Q. A. SIEG, Corydon, Harrison County, Indiana, says:

We have but few malarious or contagious diseases among farm-animals in Southern Indiana. Incident to the hog, we have what is known as cholera, quinsy, and measles. The cholera, in my judgment, is typhoid fever, and is very contagious. I have examined some hogs that died with what was called cholera. The symptoms seem, from what I have noticed, to be about as follows: First, prostration with dullness and stupidity; in most cases diarrhœa with chilliness and irregular fever. Subsequently there is an increase of the cerebral difficulties, dry skin, tenderness of the abdomen, particularly the sides of the same, an eruption of purple spots on the abdomen and thorax, and generally a cough. Usually, in eight or ten days, mortification of the bowels sets in, and the hog dies. Now and then a hog gets well, but it is an unusual occurrence. All remedies so far are failures. I would advise keeping all hogs inclosed and not permit them to run at large; then if a hog should become diseased, isolate it immediately. Before adopting this plan I lost a great many hogs, but since practicing it I have lost but very few, and what I did lose were infected from hogs lying around inclosures where mine were confined. I am therefore of the opinion that if this rule was adopted by farmers generally, that what is known as hog cholera would almost disappear.

Quinsy, I think, is the same disease as that which afflicts human beings, and requires about the same treatment. Measles never kills a hog, but if butchered while afflicted with the disease, the meat is unfit for use.

In sheep no disease except foot-rot prevails, and that only occurs in low, damp ground or by stabling in a damp, unhealthy shelter. As a remedy, the sheep should be removed to high, dry ground and separated from the well ones. Then take carbolic acid, weaken it with water, and inject the solution into the feet of the sheep every other day until a cure is effected, which usually takes from six to ten days.

Chicken cholera prevails to a greater extent than any other disease, and causes more loss. It is not confined to chickens alone, but affects ducks, geese, and turkeys alike. No remedies have been discovered that have proved of any benefit. I think good, clean, healthy apartments, with plenty of nutritious food and a good range, will greatly tend to prevent the disease. I have noticed that during the butchering season on the farm fowls are entirely free from disease, and I would infer from this that plenty of meat tends to prevent many of the maladies to which they are subject.

Mr. W. H. TROBINGER, Whitesborough, Tex., says:

Cattle are very healthy, except those imported from Missouri, Illinois, Kentucky, Ohio, &c. These are nearly all attacked with a fever within one or two months after their arrival, and at least one-half of them die. The symptoms are high fever, costive bowels, loss of appetite, and general listlessness. The duration of the disease is from one to two weeks. Remedies are various, but none very successful. Post-mortem examinations usually show signs of enlargement of liver and spleen, with inflammatory action of stomach and bowels. We have no reliable remedy.

Hogs have been very liable to disease for five or six years. Almost every disease that attacks animals of this class is pronounced cholera, but I have seen but few cases that could legitimately be thus called. The symptoms of the majority of cases that have come under my observation seem to be something like the disease called " quinsy," a swelling of the glands of the jaw. The duration of the disease is usually only a day or two. The remedies are such as calomel, scarification of the affected parts, nux vomica, and even strychnine.

We are very much troubled with disease among all kinds of fowls, which, I think, is well named cholera. The symptoms are excessive purging of the bowels and loss of appetite. They die within one or two days. The remedies are as various as the whims of men. The most successful that have come under my observation are madder, capsicum, calomel, and the mineral acids. Nitro-muriatic acid has considerable reputation as a preventive. Dose from one to two drops.

Mr. T. M. SCOTT, Melissa, Collier County, Texas, says :

All classes of farm animals are singularly free from diseases here. Occasionally a horse dies from blind staggers, brought on by carelessness in feeding unsound corn ; very rarely one is lost by colic from eating unripe corn. With the exception of the epizootic some years since we have had no prevailing diseases among horses in this county for fifteen or twenty years. Native cattle are also free from any prevailing diseases. Aged imported cattle are very apt to die within the year they are brought here; it is not known from what cause. Cattle under one year old are almost sure to live and do well.

I have heard of cholera now and then among hogs, but could never trace it to a reliable source. There has been none to my knowledge in this neighborhood for twenty-five years. In some neighborhoods cholera prevails among chickens. This is the only disease known among fowls here. No remedy is known or used.

Mr. J. A. APPLEGATE, Mount Carmel, Franklin County, Indiana, says :

The symptoms of hog cholera, so called, as given in the Agricultural Report for 1875, page 429, is better than I could give, and is an exact description of the malady which robs the country of millions of dollars annually. All other stock are exempt from any particular form of disease in this county, except fowls, which die of cholera, a disease which I am not able to properly describe, but which, on my farm, we have always counteracted by placing copperas in the drinking-troughs. It has not, however, proved a specific among my neighbors.

We regard hog cholera as very contagious and incurable. It may be prevented. It would be greatly lessened if swine were not permitted to run on the highways.

Mr. NIXON HENLEY, Monrovia, Morgan County, Indiana, says :

Among sheep we have a disease known as paper-skin, which has proved fatal in most cases. Scours is the first symptom of the disease. The sheep loses flesh and dwindles away very rapidly ; the skin becomes thin and apparently rotten—at least it is very tender. The disease is more prevalent among lambs than among older sheep. Those attacked usually die within two or three weeks. No remedy has been discovered.

The all-prevailing disease among hogs is cholera. It is very fatal, the losses being at least 50 per cent. of those attacked. No remedy has been found that will do to rely upon as a certainty.

Among chickens there are several prevailing diseases, the most prominent and fatal of which is known as cholera. Roup and gapes prevail to a limited extent.

Mr. CHARLES LARAMORE, Knox County, Indiana, says :

A few cases of hog cholera have occurred in this vicinity recently, but the disease is so well known that a description is not deemed necessary.

There have also been a few cases of a disease known as " black leg " among cattle. The animal generally becomes affected in one or the other of its legs, is very dull, and does not seem desirous of moving about. The part affected is usually swollen, and on removing the hide, after death, the diseased part presents a bloody and almost black appearance. Sometimes a fluid substance is found beneath the skin, and the flesh is then of a yellowish or pale bloody color, and presents the appearance of jelly. Animals attacked with the disease generally linger from one to three days.

A few horses have died of a disease which puzzles the horse-doctors of this locality. It has generally proved fatal. The symptoms are a loss of appetite and wasting of the flesh without any apparent show of pain. The result is generally death within from two to six weeks.

Mr. B. I. Van Court, O'Fallon, Saint Clair County, Illinois, says:

The only disease among the farm animals in our section, that gives the farmers much concern, is that affecting our hogs. There is no other disease of an uncommon character affecting at present any other class of our domestic animals. There has been some Texas fever in a few exposed localities, but nothing serious, and no spread of the contagion. The effects of the catarrhal epidemic (epizootic) among our horses is very plain in the entailment of a disease resembling in many cases bronchitis. There have been some cases of this disease, when taken in its early stages, that have yielded to the usual remedies, and where the animals have been handled with proper intelligence ; but where there has been neglect in early and prompt treatment the disease soon passed into a chronic stage, and thence from bad to worse until the lungs became involved, and a disease generated as dangerous and equally as contagious as glanders, and, indeed, very much like that fearful malady. On the discovery of those secondary symptoms the animal should be at once removed to some isolated portion of the farm, where contact with other animals would be impossible, or the animal destroyed at once, which, perhaps, would be best.

Our swine are affected by two apparently well-marked diseases. In one case the bowels are very much relaxed, and the stomach weak and unable to perform its functions. This disease is called hog cholera. It can be cured, and will yield to the usual remedies if taken in its early stages. The other disease is exactly the reverse. Instead of a relaxed state of the bowels, there is a stubborn constipation and very high fever. The animal is droopy in the early stages of the disease; it lies around in isolated places and is hard to arouse ; but at this stage it can be induced to eat, and if proper remedies are immediately administered the disease can be controlled. However, if neglected, death closes the scene in about ten days.

In 1874, my hogs (Black Berkshire breed) became affected with the disease. I had twenty head, twelve of them being about eighteen months old, and the others about ten. They were running in a wood-pasture in which there was a very heavy acorn mast that season, upon which diet they seemed to be doing well, and it was my intention to let them run there until about the middle of October and then put them up to fatten. About the tenth of October I found some of them sick. I drove them up home and gave them some corn, of which but few of them would eat. I separated the sick ones and turned the others out into the pasture again. I discovered that the hogs had high fever, and were laboring under a very costive state of the bowels. I noticed their efforts to evacuate, but with the most scanty results. Indeed, the bowels seemed almost totally obstructed. They had a hacking cough also, which is always a dangerous symptom in all hog diseases.

I concluded that the main trouble, or a very patent cause at least, was the obstructed condition of the bowels, caused by the stringency of an exclusive acorn diet. I put up one hog in the pen for treatment, after all efforts had failed in inducing them to eat.

I drenched the hog I had in the pen with common epsom salts, without any effect. I then repeated the dose, which soon produced the proper result. The animal seemed to be very weak—it could not stand upon its feet. I then gave it a decoction of Peruvian bark, calumba root and a little paregoric. This seemed to strengthen and quiet the bowels. After the animal had rested a few hours and had somewhat recovered from the effects of the severe purging, I gave it some corn-meal mixed with milk, made quite thin ; but I had to force it down its throat. The hog soon began to revive. I gave it corn-meal and milk and added a little sulphate of iron, of which it would eat. In a few days it was well, and required no more extra care. The other hogs all died. I would here state that I have had no disease among my hogs since, but I attribute it to the fact that I have given them pulverized sulphate of iron about twice a week, mixed in corn-meal or bran mash, during the fall season.

In conclusion I would state that I will always believe that the constipated condition of the bowels was a prominent predisposing cause in the development of this disease, while malarial influences were acting at the same time. There are evidences abundant, to me at least, that the malarial condition of our atmosphere, particularly in the fall season, has much to do with diseases of farm animals, in this locality at least. When the immense growth of vegetation has attained its highest degree and begins to decay, there are miasmatic conditions of the air that not only very seriously affect human life, but animal life as well ; and while we are aware that the internal organism of the hog comes nearer to that of the human than any other animal of the farm, I cannot see why they would not be affected with like conditions of the atmosphere. Permit me to call your attention to another fact which, I think, is well worth consideration, viz: There are but few, if, any, of the diseases common among our hogs found north of what may be termed the malarial line, say 43° north latitude.

Mr. T. H. Collins, Paoli, Orange County, Indiana, says:

There is no disease which affects farm animals in this section except that generally known as hog cholera, which is very prevalent. The animal affected first refuses to cat, appears restless, and after the first day becomes stupid and inclined, when let alone, to lie in its bed; is feverish, and appears to be afflicted in much the same way as a human being suffering with typhoid fever. About one-half of those affected die. The others improve very rapidly after the disease is worn out, which takes about two weeks. There is no known remedy. Some practice feeding fresh meat, which, in a few cases, is reported as having proved beneficial. A change of location has, in nearly every instance, produced a cure. I have, for the past ten years, been surrounded on all sides by the disease, yet have had no case of this or other maladies among my hogs. I attribute this to the fact that my hogs have had good care and plenty to eat, a *clean* bed and large range, and pure water to drink, with an abundance of skim-milk and other refuse from the dairy, together with a feed of refuse from the slaughter-house once a month.

Mr. M. W. WILMETH, McKinney, Collier County, Texas, says:

Among horses, we have all the old diseases known to farriers, such as botts, colic, &c.; but our most fatal local disease is known as blind staggers. An attack of this disease, on an average, lasts about twelve hours. The animals, so far as my observation enables me to judge, always die of the disease. Boring into the lower part of the forehead, between the eyes, has been tried, but without success. All other remedies have alike proved abortive. Spanish fever also prevails at times. The animal has fever, and is much affected in the loins; lingers sometimes months before dying. In some cases the disease is cured by bathing the loins in strong brine. This disease is not so common as formerly.

We have, among cattle, the common diseases known as bloody and dry murrain, both of which may be cured by purging the animal with rhubarb. This disease proves very fatal unless attended to in time, say within twenty-four hours after the attack. We have, also, what is known as Spanish fever. The animal is taken with a high fever, is much affected in the loins, and has short breathing. Cured by using a strong tea made of Jamestown weed, either of the seeds or leaves, and drenching the animal with a quart.

We have had cholera here recently among our hogs. It is a thing of late date with us, and is always fatal, as no cure has as yet been discovered. We also have among our hogs a kind of lung fever, which is very destructive. It makes its attack like Spanish fever among horses and cattle. Some cases have been cured by the use of calomel and arsenic.

Among fowls we also have what is known as cholera, though this name seems to be applied to all diseases among chickens. Alum, copperas, &c., are used with some effect.

Mr. SAMUEL WARMOTH, Princeton, Gibson County, Indiana, says:

The only animals affected with diseases in this county this year, or for several years past, are the hogs. The disease is known as cholera, and has this year carried off at least one-half the hogs in the county. Young pigs are generally the first to be attacked, and very often they all die. Then it attacks the older hogs, and, as a rule, half of them die—sometimes more and sometimes less.

The disease does not act the same in every case. Some of them are severely purged and lose their appetites and refuse to eat. Some die suddenly, while others will live for weeks moping about without eating anything. Some of them will lose a portion of their flesh, which falls off the bones while they are yet alive.

Farmers have different ways of treating the disease, but I believe there is no cure after the malady has passed a certain stage. I think it is brought on by worms, and therefore, if the worms could be kept out of the hogs they would not be liable to the disease. Salt and hickory ashes, with sulphur and copperas, will be found good preventives. Any one who will find a sure cure or preventive will deserve the thanks of the American people.

Mr. J. D. McCLANAHAN, Falmouth, Pendleton County, Kentucky, says:

The best preventive that I have tried for hog cholera is soda-ash and barilla. I give a tablespoonful of this mixture to every six hogs. The way to prepare the mixture is to put the drugs in a kettle, add two or three gallons of water, heat until the medicine is dissolved, then make bran mash with the water. One dose a week is sufficient to keep hogs in a healthy condition; however, last fall I found it necessary to give my fattening-hogs a feed of this kind every day, for the reason that some of them showed continued symptoms of disease. I lost but one out of thirty-three, and that one died on the same day that I put them up to fatten.

Mr. W. S. HAVILAND, Cynthiana, Harrison County, Kentucky, says:

In August, 1856, I lost about 30 per cent. of my hogs by cholera. I removed the remainder (about 85 head) from a blue-grass woodland pasture, supplied with a creek of running water, to a dry upland clover and timothy pasture having no water in it. They all seemed, at the time, to be more or less affected with the disease. I gave them all the corn they would eat, and regularly fed them six pounds of salt well stirred and mixed with fifty pounds of half-rotted, strong wood-ashes every seven days. They all got well, and I have never had any hogs do better than those eighty-five head did after their recovery. Of late years, while the disease is prevailing on adjoining farms to my own, I carefully notice my hogs, and when I discover lice or nits on them I wash them with soap-suds made of strong country soap, about once in ten days until all appearances of lice and nits have been removed. I then use soft soap, diluted with hot water to the thickness of thin molasses, besmear it over the head and neck of the hog, and put as much in the ears as I can, in order to drive out the lice. With this treatment I have succeeded in warding off diseases among my hogs. I salt regularly every seven days, especially in dry hot weather, giving seven pounds of salt stirred and well-mixed with about fifty-five pounds of damp, half-rotted, strong wood-ashes to each one hundred head of hogs.

Mr. O. G. BROGLE, Sparta, White County, Tennessee, says:

The only disease that has ever affected sheep in this locality is a kind of distemper, brought on I think by running on one pasture too long. In fact I am of the opinion that they are rather starved into it by having to graze over ground covered with their own filth. It is called rot by some persons, and snoffels by others. As a remedy, nothing will be found better than a change of pasture. For a severe case of snoffels a strong decoction of tobacco-juice, injected into the nostrils with a syringe, will relieve the animal very speedily. A second application is rarely necessary; but should it be, once a week will be often enough to apply it.

The most terrible and fatal disease we have to contend with is that of cholera among hogs. Its ravages are fearful during some seasons. As preventives, bluestone and copperas are used to some advantage. The preparation is made by dissolving one pound each of these ingredients in hot water, then mix a bran or meal mash and feed to one hundred head of hogs. A little salt should be added to make it palatable. If cholera prevails in a herd the above amount should be given twice a week, as it would perhaps be found as good a remedy as anything else. Once every ten days is often enough to use it as a preventive.

For lice and scab in hogs, a preparation composed of equal parts of tar and lard, rubbed on the hog, and sulphur given internally, will soon free the animal of vermin. Another remedy is to empty about one gallon of turpentine into their wallow, which will also soon free them of lice.

Cholera among fowls also prevails here. We have used alum and sulphur, both as a preventive and cure, with good success. I give one-fourth of a pound of each in dough, to every one hundred grown chickens, once in two weeks, and oftener if the disease prevails to an alarming extent. I also keep an abundance of fresh lime about their resting-places, where they can get at it without trouble.

Mr. H. GOODLANDER, Milford, Kosciusko County, Indiana, says:

Farm animals have full range here and an abundance of mast, and consequently are free from disease. However, a few hogs died last spring of quinsy. Since locating here I have induced my neighbors to use sulphur freely as a preventive of disease among stock, with apparent success. Garlic is extensively used as a preventive of chicken cholera.

Hog cholera is prevailing to some extent a few miles northeast of here, but an investigation into the causes of the disease would be attended with some expense, which the owners of the stock would not be able to bear. If the government will pay the necessary expenses the investigation will be made. Your proposed investigation deserves the highest commendation.

Mr. M. S. PULLIAM, Melissa, Collier County, Texas, says:

Farm stock of all kinds properly cared for in this vicinity are remarkably healthy. Chickens are subject to cholera. We use a handful of alum in their watering-troughs as a remedy, with apparent success.

Mr. W. P. RENDER, Point Pleasant, Ohio County, Kentucky, says:

Horses here are subject to various diseases. The first is weak eyes, which seems to be hereditary. As a preventive I would recommend more care in breeding. The second is fistula, which is an ulceration of the top of the withers, caused by a hurt or bruise of the main sinews of the neck where it joins the top of the shoulders. This disease is not necessarily fatal, though it requires a great deal of care and nursing after it has commenced running. There are various remedies. Some veterinarians burn with soft soap and whisky before the pus has formed, while others rub with turpentine and warm in with a hot iron. After the fistula has commenced running, a liniment made of May-apple root is the most effectual remedy. It should be used two or three times a day, with repeated washings with soap-suds. The third is tetanus or lock-jaw, which is a fearful disease. The horse, when taken, shows a very restless disposition; the head protrudes forward; the eyes roll back and seem to sink in the head; the hind feet are drawn under; the tail is extended, while all the muscles seem drawn to their utmost tension; some fever, with short and hurried breathing. It is caused sometimes by a hurt and at other times by overheating. Full 75 per cent. of the cases end fatally. The attack is of short duration, lasting only from two to four days. We often bleed freely from the neck vein, which, in cases caused by overheating, is sometimes effectual. In cases caused by a hurt I am of the opinion there is no remedy whatever.

With the exception of a few cases of abortion in cows, cattle and sheep are generally healthy.

Hogs are affected with a lameness which seems to be a forerunner of the cholera. They become lame in one or more of their feet; have ulcers on their joints, which last in some cases twelve months. Some have sores at every joint and finally get well and make good hogs. We have no remedy.

I have known some instances of chicken cholera where all the fowls on a farm have died. They fall from their roosts and die during the night. Like cholera in the hog, there seems to be no remedy.

The following elaborate paper on the "epizootic and enzootic diseases of swine," commonly known as "hog cholera," is from the pen of Prof. H. J. Detmers, V. S., Bellegarde, Kans.:

It is well known that some very fatal and destructive diseases of an epizootic and enzootic character have been, and are yet, prevailing among swine in several parts of the Mississippi and Missouri valleys. The farmers, not understanding the morbid proceses, and not knowing or rather not seeing the causes which produced the mischief, and finding the diseases to be very malignant and epizootic (affecting many animals at the same time), bring them all under one head and give them the rather strange and decidedly improper appellation of "hog cholera," a name which has wrought a great deal of mischief. It conveys the very erroneous idea that the disease or diseases so called must be identical with, or at least similar to, the cholera of men, consequently very contagious, and a product, not of common and local, but of very uncommon and extraordinary agencies and influences. As a natural consequence, the real causes, although near enough at hand, are overlooked and entirely disregarded, or considered as something innocent or ont of the question; and improbable, imaginary, and unknown or mysterious influences and agencies are looked upon as the possible causes. As a further consequence, almost every one who suffers losses, instead of looking the facts squarely in the face by investigating the causes, endeavors to discover specific remedies which do not exist and can never be found. Even State legislatures have offered high premiums for such a discovery. All this diverts attention from the existing facts as revealed by the morbid process and by the morbid changes found at *post-mortem* examinations, which prejudices the minds of a great many observers.

About a year ago I spent (at the request of the Missouri State Board of Agriculture) nearly a month, from August 11 to September 4, in several counties of Missouri for the purpose of investigating those diseases of swine known to the farmers as "hog cholera." I examined several hundred sick animals in the counties of Jackson, La Fayette, and Saint Charles, and made, during the time mentioned, almost daily *post-mortem* examinations, not only of hogs that had just died, but also of animals affected with disease in every stage of development, which were killed by bleeding for that special purpose. The premises on which the diseased animals had been kept were carefully examined, and the care and treatment which they had received before getting sick, and the mode and manner in which they had been kept, were ascertained by diligent inquiry and observation. Hence considerable material, sufficient to form an opinion as to the nature and real causes of the disease, or rather diseases, was collected.

Before I proceed further I wish to remark that I intend to restrict my report or communication to what I have seen and observed myself, knowing very well that still other diseases of swine, such, for instance, as various forms of anthrax, and even morbid affections caused by the presence of intestinal worms, are also called hog cholera by a great many farmers, and—one should scarcely believe it, but it is true—by a large number of agricultural papers.

Intestinal worms are a very common occurrence in an omnivorous animal like a hog, but the same, if *trachina spiralis* and *cysticercus cellulose* (the well-known bladder-worm of *tænia solium*) are excepted, seldom cause very serious damage, provided the hog is otherwise healthy, and is well kept and well fed. As to anthrax diseases, I do not think they are very frequent in the West; at any rate, I have had no occasion to observe any of the various forms of anthrax plainly developed in swine since I have lived in Kansas (nearly five years). Excluding anthrax diseases, and disorders caused by intestinal worms, I have said that *diseases* (more their own) are called hog cholera, because the symptoms of disease, and especially the morbid changes found at the *post-mortem* examinations, differ so much in different patients as to make it impossible to assign them all to one and the same disease. Still, as the morbid process is essentially the same in every case, and the differences presented are mainly due to the fact that the seat of the disease is sometimes in one organ, or set of organs, and sometimes in another, the diseases may be considered as closely related to each other, and, from a practical standpoint, it may be advisable to treat the same as members of one family, or as different forms of one and the same morbid process.

THE NATURE OF THE DISEASES.—In a majority of cases the morbid process presents itself as a catarrhal-rheumatic, and in others as a gastric-rheumatic, or bilious-rheumatic inflammation, and exhibits always, more or less plainly, a decidedly typhoid character. As a catarrhal-rheumatic inflammation it has its principal seat in the mucous membranes of the respiratory passages, in the substance of the lungs, in the pulmonal pleura, or serous membrane coating the external surface of the lobes of the lungs, in the costal pleura, or serous lining of the internal surface of the chest, in the diaphragm, and in the pericardium or serous sac inclosing the heart. As a gastric-rheumatic inflammation the principal seat of the disease is found in the abdominal cavity, but especially in the liver, in the spleen, in the large and small intestines, in the kidneys and ureters, and in the peritoneum or serous membrane lining the interior surface of the abdominal cavity, and constituting the external coat of most of the organs situated in that part of the body. The name of "hog cholera," therefore, as has been said before, is, in more than one respect, an ill-chosen one. It should be abolished at once, and a more appropriate one should take its place. As such an one I have proposed "EPIZOOTIC INFLUENZA OF SWINE," for two reasons:

First, the disease bears, in all its morbid features, and especially in the diversity of its forms, produced by the differences of the parts or organs which in different animals become the seat of the morbid process, a striking resemblance to the yet well-remembered epizootic influenza of horses, which, a few years ago, swept the whole country from the Atlantic to the Pacific. Still I do not wish to be understood as saying that the epizootic influenza of swine is identical with the epizootic influenza of horses. The resemblance, besides the epizootic spreading and the typhoid character, is limited to the symptoms and to the morbid changes. An important difference is presented by the greater malignancy of the disease of swine.

Secondly, a name derived from a conspicuous or characteristic symptom, or from an important and constant morbid change—pleuro-pneumonia of swine, for instance—might be more convenient if the main seat of the morbid process were always in the lungs and the pleura, or invariably the same in every patient; but, as the seat of the disease is not limited to the respiratory apparatus, but is also frequently formed in the parts and organs situated in the abdominal cavity, and sometimes even in the centers of the nervous system, a name should be chosen comprehensive enough to cover all the different forms in which the disease is able to make its appearance, and, at the same time, sufficiently distinct to prevent diagnostic confusion. Epizootic influenza of swine will, I think, answer the purpose.

SYMPTOMS AND MORBID CHANGES.—As the morbid process has its seat in various parts or organs of the animal body, the disease presents itself in different forms and manifests itself by different symptoms, so that, at any rate, besides other complications, two principal and two subordinate forms or varieties must be discriminated.

1. THE CATARRHAL-RHEUMATIC FORM.—This is the most frequent of the two principal forms. The morbid process has its main seat in the respiratory organs; the disease presents the features of a respiratory disorder, and either the catarrhal or the rheumatic character predominates, or both are equally developed. If the latter is the case, the whole respiratory apparatus may be found diseased. If the catarrhal character is the one most developed, the principal seat of the disease will be found in the larynx, in the windpipe, in the bronchial tubes, and, to a greater or less extent, in the substance of the lungs. If the rheumatic form is the predominating one, the principal morbid changes occur in the serous membranes of the chest (the costal and pulmonal pleura and the pericardium), and also to some extent in the tissue of the lungs. In most cases, however, the catarrhal and rheumatic character are blended with each other, and the respiratory passages, the tissue of the lungs, and the serous membranes, or portions of them, are more or less diseased. Animals affected with the catarrhal-rheumatic form indicate the presence of the disease by a short and more or less hacking cough—generally one of the first symptoms—by difficulty of breathing, a parting

or drawing motion of the flanks at each breath, by holding the head in a peculiar, stretched, and somewhat drooping position, by a slow and undecided gait, a peculiar hoarseness when caused to squeal, &c. The attending fever is severe enough to announce its presence by unmistakable symptoms, such as accelerated pulsation, changeable temperature, &c. Some of the sick animals show at the beginning of the disease a tendency to vomit, and have diarrhea, while others are more or less constipated from the first, and remain so until the disease is ready to terminate in death. If the catarrhal character is the most prevailing, but especially if the morbid process has developed principally in the throat and in the windpipe, more or less outside swelling (quinsy) will make its appearance.

At *post-mortem* examinations some important morbid changes are found invariably in the lungs. Portions of the same have become impervious to air by being gorged with exudation. The diseased tissue has lost its spongy texture—has become heavier and more morbid, and similar in consistency to a piece of liver, a condition called "hepatization." In some cases the diseased or hepatized parts of the lungs present a uniform red or reddish-brown color, an indication that the exudation has been produced and deposited in the tissue of all the diseased lobules at the same time, or without interruption. In other cases the diseased portions of the lungs present different colors ; some are red, some brown, and others gray or yellowish-gray, which gives the whole hepatized part a somewhat marbled appearance, and shows that the exudation has been produced and deposited at different periods. The gray hepatization, which, in such a case, is the oldest, and the brown, which comes next in age, frequently contain a few tubercles, and even here and there a small ulcer interspersed. Otherwise neither ulceration nor suppuration has been observed. Important morbid changes are usually also formed in the serous membranes of the thorax. The same consist in a more or less firm coalescence between parts of the pulmonal pleura and the corresponding parts of the costal pleura, or of the diaphragm, and in an accumulation of a larger or smaller quantity of straw-colored water or serum in the chest. In some cases, especially those in which the rheumatic character has been very predominating, the morbid products of the diseased serous membranes are frequently very copious ; the adhesion between the costal and pulmonal pleura, or between the internal surface of the walls of the thorax and the external surface of the lungs, is usually very extensive, and parts of the posterior surface of the lungs are sometimes found firmly united with the corresponding surface of the diaphragm, or membranous partition between the chest and the abdominal cavity. The quantity of serous exudation or straw-colored water deposited in the chest is often very large, and the pericardium, too, in most cases, contains a larger or smaller quantity, sometimes enough to interfere seriously with the functions of the heart, and to constitute in that way the immediate cause of death. The blood is found to be thin and watery in every case, coagulating rapidly to a uniform but rather pale-red clot of a loose texture. Its quantity is always very small.

2. THE GASTRIC-RHEUMATIC FORM.—This form presents itself not quite so often as the catarrhal rheumatic, but is fully as malignant, and constitutes the second main form which the disease is found to assume. The morbid process has its principal seat and produces the most important morbid changes in some of the organs situated in the abdominal cavity, but especially in the liver, in the spleen or milt, in the kidneys and ureters, in the intestines or guts, and almost invariably in the peritoneum or serous membrane which lines the interior surface of the abdominal cavity and constitutes the external coat of nearly every intestine.

The symptoms which present themselves while the animal is living differ not very conspicuously from those observed in the catarrhal-rheumatic form. The short, hacking cough, characteristic of the latter, is more or less wanting ; the difficulty of breathing is less plain ; the weakness in the hindquarters, and the staggering or unsteady gait observed only in a limited degree in the catarrhal-rheumatic form are more conspicuous, and the fever is fully as high in one form as in the other. In some cases the affected animals arch their backs, or rather the lumbal portion of the same, to a very high degree, and form an outline similar in shape to an ∼. I observed this especially in such cases as those in which the seat of the disease was found to be in the kidneys and in the ureters, and in which a large quantity of serum or water had accumulated in the abdominal cavity. Animals affected with the gastric or bilious-rheumatic form are usually more or less constipated. The dung, which is voided in form of small, irregular-shaped balls or lumps, is often coated with a layer of grayish or discolored mucus, and has the consistency of shoemaker's wax. Toward the end, however—that is, if the disease has a fatal termination—the costiveness usually disappears, and is followed by a profuse and very fetid diarrhea, which may be looked upon in every instance as a forerunner of death.

The principal morbid changes, as I have found them, are as follows:

1. Degeneration of the liver, brought about by a copious exudation infiltrated into the tissue of that organ. Such a degeneration, although not a constant morbid change, is found quite often. In some (not very frequent) cases a few tubercles, and in others

(still less frequent) even a few very small abscesses have been found imbedded in the diseased substance of the liver.

2. Morbid enlargement of the spleen or milt. I found this change in nearly every case. In some cases the enlargement was not very conspicuous, but in others the spleen was more than three times its natural size, was perfectly gorged with bl'od, presented a dark or black-brown color, and was so soft that a very slight pressure with a finger was sufficient to sever its tissue.

3. In quite a large number of cases I found either one or both kidneys diseased and enlarged, and presenting an inflamed appearance. In one case both kidneys and both ureters exhibited a high degree of inflammation and considerable gangrenous destruction. The latter, however, was probably not a consequence of the disease; the animal had been drenched repeatedly with oil of turpentine, and was the only one in which I found any gangrene. In another animal, which, by the way, was already convalescent, and was killed by bleeding, I found one kidney enlarged to three times its natural size, its pelvis very much distended, and its funnel-shaped ureter dilated to such an extent (where it proceeds from the kidney) as to present a diameter of nearly one inch and a half. The walls of the ureter were very thick and callous, especially at its anterior, funnel-shaped end, and the latter contained in its interior a semi-solid fibrous substance, which occupied the whole cavity, and extended even into the kidney.

4. In some cases I found the membranes of the intestines or guts, but especially those of the jejunum or small intestines, the coecum, and colon or larger intestines, and also of the rectum, in a more or less inflamed and degenerated condition. In two cases a whole convolution of the jejunum had united to an almost solid bunch. On opening the latter I found in each case all three membranes, but particularly the external or serous membrane and the internal or mucous membrane very much swollen and degenerated, the passage nearly closed, and in a small cavity in the center of the bunch one (in one case) and two (in the other) large round worms (apparently *Echinorhynchus gigas*) imbedded. In another case I found, besides other morbid changes, a few round worms in the stomach, and in the mucous membrane of the guts or intestines a large number of callous scars, such as are usually left behind where the gigantic *Echinorhynchus* or hook-headed worm has been fastening itself. These three cases are the only ones in which I have found any entozoa or worms in the digestive canal.

5. In almost every case I found larger or smaller portions of the peritoneum or serous membrane, which lines the inner surface of the walls of the abdominal cavity and the external surface of nearly every intestine, swelled, more or less inflamed, and morbidly changed. In some cases even a coalescence between parts of the intestines, especially of jejunum and rectum, and the walls of the abdominal cavity had been affected. In one case a part of the jejunum had become firmly united to the lower border of the right lobe of the liver, and in another the whole rectum adhered so firmly to the upper wall of the pelvis and of the posterior part of the abdominal cavity, that it required the use of the knife to effect a separation.

6. In every animal that had been affected with the gastric-rheumatic form I found a larger or smaller quantity of straw-colored water or serum, and small lumps and flakes of coagulated fibrine in the abdominal cavity; in some cases the quantity was quite large, and in others comparatively small.

As subordinate or complicated forms, I look upon such cases in which either one of the principal forms—the catarrhal-rheumatic or the gastric-rheumatic—is essentially modified by being complicated with an affection of the brain and its membranes, or with a serious disorder of the lymphatic system. Two subordinate forms, therefore, must be added.

3. THE CEREBRO-RHEUMATIC FORM.—The same, though always blended with and to a certain extent subordinate to, one of the principal forms, has been observed in a large number of sick animals. The latter, besides exhibiting all the symptoms of one or another of the two principal forms, shows also plain indications of a morbid affection of the brain. The same consists principally in partial or perfect blindness, a very staggering gait, and aimless movements in general. On opening the skull I invariably found more or less swelling in the membranes enveloping the brain, a larger or smaller quantity of serum deposited inside of the dura mater (hard or external membrane), the substance of the brain more or less softened, and the ventricles (small cavities) of the brain filled with serum. The other morbid changes found did not differ from those described under the head of catarrhal-rheumatic or gastric-rheumatic forms respectively.

4. THE LYMPHATIC RHEUMATIC FORM.—The same, too, has been observed quite often, but always as a complication of one of the principal forms described—subdivisions 1 and 2. The whole morbid process presents a somewhat scrofulous character. The lymphatic system is plainly affected; tumors and ulcers, showing a scrofulous character, are found in various parts of the body, but especially on the gums. Hence there can be no doubt that such cases, although complicated and blended invariably to such an

extent with one or another of the main or principal forms as to make it impossible to draw distinct lines, have to be looked upon as a subordinate form with a lymphatic character. I have been informed repeatedly by reliable persons that in some of the sick animals cutaneous eruptions have constituted one of the most conspicuous symptoms of the disease. If this is a fact, it is possible that yet a fifth form has to be added—*erysipelatous*. Still I had no chance to examine such a patient, notwithstanding I have examined a large number of sick animals, exceeding, I should judge, one thousand. I am, therefore, not prepared to decide whether the cutaneous eruption is a product of the same causes or influences which are at the bottom of the other morbid changes, or whether the same is an independent disease, and merely an accidental complication.

It is probably not necessary to mention that all the morbid changes which have been described as the products or attendants of a certain form are but seldom found as a total in one and the same animal, as some of them are either usually missing or but little developed. Neither will it be essential to state that even the two principal forms of epizootic influenza of swine, leaving the subordinate forms out of consideration, are scarcely ever observed entirely independent of each other or without being in the least complicated with each other; that, on the contrary, the gastric-rheumatic and the catarrhal-rheumatic are not seldom blended with each other to such an extent as to make it very difficult to decide which one has to be considered as the most predominating. In each case the symptoms, too, are blended with each other, and morbid changes, frequently of equal importance, are found in both large cavities, in the chest and in the abdomen. These facts are easily understood by any one who is at all familiar with pathology and with morbid anatomy. The main or predominant character of epizootic influenza of swine is always rheumatic, and the principal seat is in the system of serous membranes which abound in every large cavity of the animal body. Serous membranes not only line the interior of those cavities, but constitute also the external coat of nearly every internal organ. Hence it is but natural that such a disease localizes itself in many different parts of the animal organism, produces in consequence different morbid symptoms, and causes different forms of disease. It is true that in some cases the disease exhibits a prevailing catarrhal character, but if it is taken into consideration that the causes of rheumatic affections and of catarrhal diseases are often essentially the same, and that not only the seat but the character of the disorder depends frequently upon an individual predisposition of the animal, a further explanation will not be needed.

THE CAUSES—To ascertain the causes has been my principal object. It was, therefore, necessary to observe a large number of cases, and to investigate the disease in different localities. This I have done, and have come to the conclusion that at least some of the causes, and I think I make no mistake if I say the most important ones, are of such a nature as to admit removal, notwithstanding they are diverse and numerous, and have their source, to a certain extent, in the manner of farming and stock-raising, or rather hog-raising, customary in the West. Although I will not deny the possibility of an existence of certain agencies of a miasmatic character, nor the possibility of a presence of a micrococci or other microscopic sporules calculated to act as a cause or to contribute in producing the disease, I must confess that if anything of that kind has been acting as a cause, it has escaped my notice. In the first place I had no microscope at my disposal, and secondly I have not been able to discover anything in the whole morbid process nor any morbid change that cannot be the product of those noxious influences which I consider as the main, if not exclusive, causes of the disease, and which in my opinion are well able to produce every one of those morbid changes which I had an opportunity to observe. Those injurious influences or agencies which I am obliged to consider as the principal causes act in different ways, and, for a better survey, may be divided into three classes. As belonging to the first class I look upon everything that will interrupt or disturb the perspiration. In the second class I place all such noxious influences and agencies as interfere directly with the process of respiration. Finally, in the third class I put all such noxious agencies or injurious influences as tend to aggravate the disease if already existing, by aiding in making its character more typhoid, or which produce a special predisposition, by weakening the constitution of the animal.

1. *Injurious influences which act as a cause by producing an interruption or partial cessation of the perspiration.*—These influences are numerous, and of much greater importance than one who looks at them superficially may be inclined to suppose. The skin of an animal is a very important organ ; it not only serves as a protecting tegument, but has also other vital offices which are scarcely of less importance to the existence of the animal organism than those of the lungs. The skin discharges through its pores a large amount of wasted material, and absorbs aeriform and liquid substances from the outside world. Consequently, it may be looked upon as an organ whose duty it is to supplement the functions of several other organs, but especially those of the lungs and of the kidneys. To ascertain the effect which a total interruption of the functions of the skin would have upon the animal organism, interesting experiments have been made by Bouley,

Magendie, Gerlach, and others. A complete interruption was brought about by covering the skin of various animals with an air-tight coat of varnish, grease, or tar, and the result, according to Gerlach, was as follows: Accelerated pulsation, extraordinary fullness of the arteries till an increased discharge of urine made its appearance, somewhat accelerated breathing, trembling of the whole body, rapid emaciation, great debility, augmented secretion of an albuminous urine, which latter contained also some of the coloring matter of the gall, and a decrease of the animal temperature. The latter, however, became not very conspicuous before the animal had become emaciated, and was near dying. The animals (horses) so treated died within from three to ten days. Pigs smeared all over with grease or fish-oil, for the purpose of killing lice, died within a week, and showed the same symptoms.

The office of the skin, at least as far as the processes of elimination and absorption are concerned, bears also a very close relation to the functions of the diverse serous and mucous membranes. It is true if the skin is disqualified to perform its allotted duties, or if its functions are interrupted by some means, the same will be performed partially but partially only by those organs named the lungs, the kidneys, and the serous and mucous membranes in general. These organs, in such a case, have to make extraordinary efforts if the equilibrium in the organic change of matter, so indispensable to the preservation of health, is to be maintained only approximately. To maintain a perfect equilibrium is impossible, for these organs, as I have said, can, in addition to their own duties, only partially perform the functions of the skin; certain parts of wasted material will not be discharged, but will remain in the organism. The lungs, the kidneys, and the serous and mucous membranes, if I may use the expression, will be overburdened, and the consequence will be that just those organs thus weakened will be the first ones that become diseased, or have to suffer from over exertion, and from the injurious effects necessarily produced by a retention of wasted matter in the organism, and also by a constant loss of organic compounds that cannot be spared. That such loss is taking place has been proven by the experiments of Professor Gerlach, which shows that the urine in such a case carries off albumen. Further, that an interruption of the perspiration must necessarily produce a disturbance in the circulation of the blood, which results in an extraordinary flow of blood to those organs—lungs, kidneys, &c.—burdened with increased activity, and constitutes in that way a cause of congestion and subsequent inflammation, is too evident to need any further explanation.

The perspiration can be interrupted, or, in other words, the skin can be disqualified to perform its functions by several means; for instance, by a disturbance or a partial interruption of the circulation of the blood in the capillary vessels; by congestion and inflammation; by any degeneration or morbid change of its tissue, or of a part of its tissue; by a closing of its pores in a mechanical way, &c. This granted, it remains to be ascertained if those pigs and hogs which are, or have been, affected with the epizootic influenza of swine (erroneously called hog-cholera) have been subjected to one or more of those just named influences or agencies able to cause an interruption or partial cessation of the activity of the skin (perceptible and imperceptible perspiration). Taking the facts just as they have presented themselves, that question must be answered in the affirmative. My investigations and my inquiries have convinced me that in all those pigs or hogs which have suffered from or died of that disease, one or more of those causes have been at work, as I shall endeavor to show.

1. All animals affected with the epizootic influenza—at any rate all those which I have seen, and I have seen a large number—were very lousy. Lice irritate the skin, keep it in a semi-inflamed condition, cause swelling, and, finally, a gradual degeneration of its external layer—beyond a doubt constitute to some extent a disturbance of normal perspiration.

2. All the hogs and pigs which had contracted the disease had been exposed night and day to all the sudden changes of temperature and weather so frequent in the Western States. Some of the animals had been kept in small, wet, and dirty yards and enclosures, without a roof to protect them; they had to suffer during the day from the rays of the sun, and from the heat which naturally accumulates in a small space or lot walled in by a tight fence, and which is constantly increased by the decomposition of wet manure and other organic substances. During the night the same were exposed to the chilling influence of the cold night air, and the frequently very heavy dews, not to mention the effect of severe rains and thunder-storms. Further, after each rain the animals thus kept had a chance to get the entire body covered with mud and the pores of the skin thoroughly closed; but an opportunity to get rid of the dirt by taking a bath was never given. Such influences, evidently, are very apt to cause irregularities in the circulation of the blood in the capillary vessels of the skin, and an interruption of the perspiration. Other animals have been kept in comparatively large herds, and have been allowed to run at large in a barn-yard, in a so-called hog-lot, in the woods, &c. These, too, were exposed more or less to the burning rays of the sun during the day, and at night, in most cases, they found shelter under a corn-crib, under an old stable, or an old barn—at any rate in the closest and dirtiest places, where they packed room, and where they often crowded on top of each other when retiring to

sleep. As a consequence the animals became heated, and, perspiring, as they left their lair in the morning took cold on coming in contact with the chilled atmosphere. A sudden cooling, however, or a sudden reduction of the temperature of the surface of the body is apt to effect a contraction of the capillary vessels of the skin, hence a diminished supply of blood, and, in consequence, a decrease or partial interruption of the functions of the skin.

The animals, thus suddenly cooled by the morning air and the wet dew, become, in the course of the forenoon, again exposed to the rays of the sun and the heat of the day, which induces them to go to the first pool of water, if one is accessible, to take a bath. This is all right and well enough, because in the summer a hog should have access to water and an opportunity to take a bath as often as it desires. In all those places, however, in which the disease had made its appearance, I found the water to which the hogs had access almost invariably so shallow and of such a limited quantity that the bathing and wallowing of one or more animals was sufficient to convert the same into sticky, semi-fluid mud. Consequently, if the herd was a large one but a very few animals—and those invariably the stronger and most active ones—had now and then a chance to find clear water, and to reap real benefit from taking a bath. All others, especially the younger and smaller animals, were compelled to wait until the first comers were through with their bathing and had changed the water to mud; the former, therefore, had scarcely ever an opportunity to clean themselves from the mud of the preceding day, and to open the pores of their skin by taking a bath in clean water. If they wished to take a little cooling they had to be satisfied with taking a mud-bath, and as every new bath was a mud-bath again the pores of the skin, instead of being opened, became closed more and more effectually from day to day, until finally the perspiration was thoroughly interrupted, and disease made its appearance as the natural result.

It is different where the herd is a small one, for then nearly every animal will sometimes have a chance to open the pores of its skin in tolerably clear water, and the perspiration will not be seriously interrupted. That these deductions must be correct is proved by the fact that in every large herd nearly all the younger and weaker animals (shoats) have become a prey to the disease, while the larger or stronger and most active animals, which are usually the first ones to go to the water in the morning while it is measurably clear, have either remained exempt or have contracted the disease in a mild form, and have mostly recovered. Finally, small herds have either suffered fewer losses, have been less severely attacked, or have remained exempted altogether. The injurious effect produced upon the system of the animal by the muddy and filthy condition of the water, which most animals so situated have been compelled to drink, will be explained hereafter.

2. *Agencies and influences which interfere directly with the process of breathing.*—These, too, as already indicated, are of different nature. When I first commenced my investigation it struck me that all those swine—pigs, shoats, and grown hogs of every age and description—which run at large in the streets and thoroughfares of Kansas City, Westport, Independence, Lexington, and other towns and villages, and lead the most independent life possible, but do not congregate—go home in the evening, and belong to persons who own but one, two, and maybe three animals; as also all those swine which are kept by themselves, either one by one or only a few together; and, finally, all those which are kept in comparatively small herds in pastures, orchards, or woods, coated everywhere with grass and perfectly destitute of dusty, bare ground and old manure-heaps, are and have been, with rare exceptions, perfectly healthy. I say with "rare exceptions," for it has been reported to me that a few of those swine running at large in the streets have died, but I have not been able to ascertain the causes of their death.

On the other hand, the herds which have been kept in yards, pastures, fields, &c, consisting partially or wholly of bare, dusty ground, or containing heaps and accumulations of old manure, have and are suffering severely, and the more according to the size of the herd and the worse the dust of soil and old manure. In large herds, composed of one hundred head or more, the mortality has been as high as from 70 to 90 per cent.; in smaller herds from 25 to 60 per cent., and where only a few animals were kept together, and consequently each animal was only compelled to inhale the dust kicked up by itself and occasionally by one or two others, the mortality has been comparatively low—has seldom exceeded 10 per cent., or fatal cases have not occurred at all. Further, in all those cases in which the hogs or pigs had been compelled to inhale with nearly every breath a large quantity of soil and manures, ground to powder by rolling, tramping, and the rays of the sun, all the *post-mortem* examinations revealed as principal morbid changes a morbid affection of the eyes, inflammation of the respiratory passages (throat, windpipe, bronchial tubes), hepatization of the lungs in various stages of development, and, in some cases, even tubercles or a few small abscesses in the pulmonal tissue, while the serous membranes (costal and pulmonal pleura, pericardium, and peritoneum) presented a comparatively healthy condition, except in those cases in which the causes described in subdivision 1 had been acting together with those under discussion. If these facts are duly taken into consideration, scarcely any doubt can remain as to the constant inhalation of powdered soil and manure constituting one of the principal causes of the epizootic influenza of swine.

As another noxious influence tending to interfere with the process of respiration, or injuring the respiratory organs, may be considered the gases or effluvia emanating from old decomposing manure heaps and from dirty and filthy pig-sties and hog-yards. Still, I must look upon them as something of subordinate importance—not *per se*, but compared with the more substantial agencies—and, therefore, do not deem it necessary to enter into further details.

3. *Causes which weaken the constitution, produce predisposition, and develop or promote the typhoid character of the disease.*—As such have to be mentioned: 1. Foul and impure water for drinking. As a general rule, hogs are usually compelled to drink either out of a dirty trough, if confined in a sty, or from muddy pools and wallows, if kept in pasture, &c., and, therefore, are frequently obliged to drink water that is not only muddy and impure, but even stinking and full of decomposing organic substances. That such water is apt to develop microscopic animal and vegetable growth, is often inhabited by the brood or the larvæ of various species of intestinal worms, and thus prepared to convey numerous germs and causes of disease to the animal organism—maybe more than are introduced in any other way—is a well known fact, and does not need any explanation. 2. The filth and manure that is consumed with the food. On almost every Western farm (at any rate on all those on which I found the disease) the swine are fed with corn in the ear; the ears of corn are thrown into the pig-sty, yard, or feeding-lot, as the case may be, but always in a place full of manure and dirt, either wet or dry. As a consequence, the animals can scarcely pick up a kernel of corn that is not soiled with filth, and are obliged to consume a great deal of nastiness. That such wholesale consumption of filth and excrements must finally undermine the constitution of even the healthiest animal, and must give to any disease that may happen to exist or to appear a typhoid character, is self-evident. 3. On a great many farms in the West the corn-cribs are either insufficiently covered or not covered at all, and, as a consequence, a great deal of the corn fed in the spring and during the summer is moldy and rotten. Moldy corn does not constitute healthy food; on the contrary, it is poisonous if consumed in large quantities; at any rate, it weakens the constitution, promotes and produces disease, especially of the respiratory organs and of the kidneys, and is well calculated to give any disease a decidedly typhoid character. 4. One very common mistake in feeding may also be mentioned as perhaps not entirely without influence. I refer to the practice of feeding nothing but corn. It may suffice, however, to say that corn does not contain in a due proportion all the elements necessary to the growth and development of an animal; it is destitute of some and contains others in too small a proportion. Hence a variety of food is just as necessary to a hog as to any other animal, if health and vigor are to be preserved. To enter into particulars would lead too far.

One may ask, if the causes of the disease are of such an ordinary character, how can it be possible that it has become such an extensive epizooty? The answer is not difficult. A satisfactory explanation can be given. 1. Notwithstanding the most diligent search and patient inquiry, I have not been able to discover any injurious influences or agencies, in addition to those enumerated, that have acted upon all of the diseased animals, or upon a large number of the same, which can be taken into consideration as possible causes. 2. The treatment or keeping of swine is essentially the same almost everywhere in all the Western States. The causes mentioned are, therefore, sufficiently discriminated or general enough to produce an epizootic. A great many farmers, who are frequently careless enough in the treatment of even their horses and cattle, usually think that a hog is but a "hog," and it can get along with "hoggish" treatment—that it delights in nastiness, filth, and dirt of every description, and does not need a dry, comfortable, and clean resting-place during the night, clean and sound food, clean and fresh water for drinking and bathing, nor shade and shelter against the burning rays of a Western sun, against wet and cold and the sudden changes of weather and temperature in general. But they are very much mistaken; there is probably no animal which repays good care and rational treatment more than the hog. Still, if nature had not endowed the same with such an excellent constitution, pork might have become, before this, a very rare article.

Some one may say, "If the principal causes of the disease have their source in the manner in which the swine are raised, kept, and provided for, which does not differ essentially from former years, how does it happen, or how can it be explained, that the disease made its appearance as an epizooty only a few years ago, and not before?" While the country was new hogs were not so numerous as now, or at any rate were not kept in such large herds; pig-sties, hog-lots, and swine-pastures contained not so much accumulated filth and manure, nor so many bare and dusty places as they do now. In the course of many years the excrements and other decomposing organic substances have not only accumulated on the surface of the premises where hogs are kept, but the ground and water have also become impregnated with the same. The disease, I do not doubt, will still spread and increase in malignancy in the same proportion in which dung and dirt are allowed to accumulate, and in which the size of the herd is increased.

Is the epizootic influenza of swine a contagious disease?—To tell the truth, I am not prepared to decide that question, because such a decision requires numerous experiments, and these I have not been able to make. A great many farmers believe, nay, hold themselves convinced, that the disease must be contagious, and have furnished me with facts which I admit seem to point very strongly that way. Still I think the epizootic character or the fearful spreading of the disease can be satisfactorily explained without the existence of a contagion. The fact that the hogs and pigs running at large in the streets of the towns and cities are, with rare exceptions, healthy and remain exempted from the disease, notwithstanding they are much more exposed to contagions or contagious infection than any others, goes far to show that the disease is probably not contagious.

Duration of the morbid process.—In some cases the disease has had a fatal termination within two days after the first plain symptoms of sickness had made their appearance, and a few cases have been reported to me in which the animals have died within from six to twelve hours; but as to the latter cases, I am inclined to think the first symptoms have escaped observation; a very common occurrence in diseases of swine. The average duration of the disease may be set down as from five to fifteen days. Still some animals have been sick from three to six weeks, but as most of these recovered, a part of that time should be looked upon as belonging to the stage of convalescence, or, if the patients died, the disease was protracted by relapse.

Prevention.—The measures of prevention consist in removing the causes or in treating the swine in a rational manner in accordance with hygienic principles. If this is done, no other special treatment nor any medicines will be needed to ward off the disease. To give medicine to healthy animals for the purpose of preserving their health is a bad practice and may be fraught with injury. The use of medicines can have but few objects, viz., to mitigate, to remove, to destroy, or to divert injurious influences. To give the same for any other purpose will do much more damage than good, and should never be done. Hence I have to caution every farmer against the use of any patent nostrums or quack medicines advertised as "cure-alls," but intended only to draw the money out of the pockets of the credulous.

But to the point: I am confident that the epizootic influenza of swine, or the disease commonly called hog-cholera, will cease to exist, or, at any rate, will lose its epizootic character and become a very rare occurrence, first, if large herds of swine are divided into smaller ones containing only a few (three or four) animals each; second, if each lot, consisting of a few animals, is provided with a comfortable pen or sufficiently-protected resting-place to sleep in, which is kept free from filth, dust, and manure, is well ventilated, and has a good roof; third, if every hog or pig has access several times a day, or as often as weather or temperature and circumstances require, to fresh and clean water for drinking and bathing, either in troughs made for that purpose or in a brook or streamlet; fourth, if no filth, manure, and other decomposing organic substances are allowed to accumulate in any of the sties, yards, pens, hog-lots, or pastures in which the hogs or pigs are kept; and, fifth, if the food is always healthy and sound and never soiled with filth and manure. I know very well that many farmers prefer to be sent to the drug store for medicine in preference to complying with these rules, and some of them may even think or say, "If I cannot keep my hogs in the old 'hoggish' fashion, but must treat them even better than I am in the habit of treating my horses and cattle, I prefer not to keep them at all." To such men I have to say, if you do not keep any hogs you certainly will not lose any, and may thus benefit yourself and your neighbor, who will reap the profit from the scarcity of hogs produced. But I can assure you that any one who will consent to treat his swine in a rational manner, as an animal ought to be treated, will gain thereby, and will receive ample compensation for his care and labor. At any rate, it will pay much better for any one to raise, for instance, fifty hogs, to keep them well and lose none and to develop them into "prime porkers" or so-called "Philadelphia" hogs, than to raise one or two hundred in "hoggish" fashion, lose from 50 to 70 per cent., and produce animals that figure as inferior "light-weights" or "scalawags" in the market reports. Moreover, the amount of food which is needed to produce two hundred pounds of inferior and frequently unhealthy pork, if the pigs are kept on the manure-heap, in the barn-yard, or in small, nasty pens, will easily produce three hundred pounds of good, healthy, and palatable pork if the keeping of the animal is always in strict accordance with the laws of hygiene. If the latter are never violated, I am sure epizootic influenza will not make its appearance; but if the indifferent, or rather negligent, treatment of swine customary in the West does not undergo a thorough change, the disease will increase in frequency from year to year.

In thus giving my views candidly and in plain language, I wish to state, without any apologies, that my object is not to blame any one, but to tell the honest truth, and to point out the way which must be pursued if it is desired to get rid of the disease. The mistakes made are not committed by a few farmers and hog-raisers, but by a great many. If those who find themselves guilty of having neglected their hogs, or of having treated them in "hoggish fashion," will accept what I have said in the same spirit in which it is given, and follow my advice, they will have no cause to regret it.

Treatment.—The treatment may be divided into two parts: a hygienic and a medical. The former, which includes a removing of the causes, is in this, like in most other causes, of the very greatest importance. If the causes are promptly removed, a great many sick animals not already too far gone may be saved. If the same are not, the very best medical treatment will be of little avail. The sick animals must be separated from the herd, must be provided with a clean and dry resting-place, must have pure air to breathe, clean water to drink, and healthy, clean, and easily-digestible food to eat.

As to the medical part of the treatment: I would recommend giving to each patient at the beginning of the disease a good emetic, composed either of powdered white hellebore (*veratrum album*) or of tartar-emetic, in a dose of about one grain for each month the sick animal is old, provided the latter is not over fifteen or sixteen grains. The largest dose to be given a full-grown animal should not exceed fifteen or sixteen grains. The emetic is best administered by mixing the same with a piece of boiled potato, or, if the hellebore (which I prefer) is chosen, by strewing the powder on the surface of a small quantity of milk, as neither boiled potato nor milk will be refused by any hog unless the animal is very sick, and in that case it will be too late to make use of an emetic. After the desired action has been produced the animal will appear to be very sick, and will try to hide itself in a dark corner; but two or three hours later it will make its appearance again, and will be willing to take a little choice food, such as a few boiled potatoes, a little milk, &c. At this time it will be advisable to again give a small dose of medicine, either a few grains (two or three to a full-grown animal and to a pig in proportion) of tartar-emetic or of calomel. Mix with a piece of boiled potato, or, if the symptoms should not have returned, mix with a small pinch of flour and a few drops of water (sufficient to make a stiff dough) and form into small round pills. I wish to remark here that a sick hog should not be drenched with medicine under any circumstances, for a drench, given by force, is very apt to pass down the windpipe into the lungs as soon as the animal squeals, and frequently causes instant death. The tartar-emetic has to be chosen if the disease has its principal seat in the respiratory organs or presents itself in its catarrhal-rheumatic form, and the calomel deserves preference if the gastric or bilious-rheumatic form is prevailing, but especially if the liver is seriously affected. Either medicine may be given in such small doses as mentioned three times a day for several days in succession, or until a change for the better becomes apparent. It is also advisable, particularly if the disease exhibits a very typhoid character, to now and then mix for each animal a few drops of carbolic acid with the water for drinking or with the slops. Convalescent animals, which have become very weak and emaciated, will be benefited by giving them once a day from a few grains to half a drachm of sulphate of iron (copperas) mixed with their food, but the use of iron must be discontinued if the patients become constipated or if the excrements turn black. Those convalescents in which the lungs have become hepatized to a considerable extent may receive repeatedly small doses of carbonate of potash for the purpose of promoting the absorption of the exudations deposited in the tissue of the lungs. The size of the dose of carbonate of potash, as well as of iron, depends upon the size and the age of the animal.

A local or external treatment is also of considerable importance. A good counter-irritant, or blister composed of cantharides, or Spanish flies, and oil, made by boiling one ounce of the former and four ounces of the latter for half an hour over a moderate fire, or for one hour in a water-bath, should be applied on both sides the chest in all such cases in which the organs situated in that cavity are seriously affected. Such a counter-irritant has usually a very beneficial result. In most cases one application will prove sufficient to relieve the animal to a considerable extent, provided the oil is thoroughly rubbed in before the disease has made too much headway, or before the vitality of the organism has been destroyed. If the effect of the fly-blister proves insufficient it may be applied again the next day, but if the same produces no effect at all it may be taken as an indication that the animal is going to die, and that any further treatment will prove of no avail. Fontenels and seatons have really the same effect as a fly-blister, but they act slower, are less reliable, and may otherwise cause damage, especially if the typhoid character of the disease is very much developed, by weakening unnecessarily the constitution of the patient.

In conclusion, I will mention that epizootic influenza of swine, or so-called hog-cholera, is not a new disease, nor peculiar to our country, as people seem to believe. It has been known in Europe for many years. Professor Spinola gives a description of an epizootic "pleuro-peripneumonia," corresponding almost exactly to the catarrhal-rheumatic form of epizootic influenza of swine, in his "Die Krankheiten der Schweine" (Diseases of Swine), Berlin, 1842, page 82 *et seq.* Another brief description will be found in the Austrian "Vierteljahresschrift fuer wissenschaftliche Veterinaerkunde" (Quarterly for Scientific Veterinary Science), Vienna, 1870, vol. xxxiii, part 2, page 137, copied from "Il Medico Veterinario," 1869, page 529.

Prof. E. F. RIPLEY, V. S., Portland, Cumberland County, Maine, says:

In regard to the diseases of farm animals I am happy to inform you that we have had no epizootic maladies the past two years. I have had quite a number of cases of pneumonia (of a low typhoid character) among horses, but the majority of the animals recovered. I treated them with stimulants, carbonate of ammonia, camphor, and capsicum. Occasionally I have a case that will bear a sedative. To some affected with extreme nervous prostration I gave assafetida and ergot. I have successfully treated fifty-odd cases of spinal and cerebro-spinal meningitis, mostly the former, where there were no brain complications. Treatment, one-half ounce of aloes, two drachms of carbonate of ammonia in bolus, followed with extract of belladonna and ergot and bromide of potassium. Some I treated with stimulants, applied mustard to the spine, and supported those with slings that were not able to rise without help. I treated six others (more severe cases, some of them down and unable to stand) that died. I have treated twenty horses suffering with pupa homeragicu. I gave them chloride of potassium, iron, quinine, and matico, mixed with one gallon of milk and six eggs, administered once a day. Most of them would drink with avidity. With an abundance of pure air and good nursing they soon recovered. We have more or less sore throat here among horses during the fall and spring seasons. Some neglected cases run into glanders.

Diseases among horned-cattle increase as the country grows older. More especially is this the case among milch-cows. Puerperal fever is the most common disease among this class of stock, and proved fatal in more than half the cases reported. Some cows die within an hour after the first symptoms of the disease are observable. I successfully treated six cases this season by giving one pound of sulphate of magnesia, twenty drops of croton oil, two drachms of Jamaica ginger, in three pints of warm water. Their milk and urine should be drawn, and mustard applied to the spine. If injections of physic do not act in six hours give half-pound doses of magnesia and ginger every six hours until the bowels move. Within two hours from the first cathartic give two ounces of spirits of nitre and four ounces of acetate of ammonia. Repeat every six hours until the animal is able to rise. I have had many cases of congestive fevers in cows and oxen, most of which have recovered. I give one pound sulphate of magnesia, two drachms ginger, and one drachm tincture of aconite. If the bowels are constipated, after the fever subsides, I give half-drachm doses of nux vomica.

I have successfully treated a few cases of entritic fever in swine with calomel and muriate of ammonia, alternated with belladonna.

Mr. S. H. LOGAN, Greensburg, Decatur County, Indiana, says:

The disease known as hog-cholera is now and has been prevalent for several years in this county. I was a feeder of hogs for several years in distilleries in Cincinnati and at Lawrenceburg, in this State. The average loss of hogs by death from this disease at distilleries I think is fully one-half. The loss among hogs in the country from the same cause, of one year old and over, will average about the same; those of six months old and under about all die, or perhaps one out of ten may live.

The remedies used here are sulphur, copperas, black antimony, saltpeter, and assafetida. These remedies have been given separately and in different combinations. Several patent medicines have also been used, but I have never known any benefit derived from any of them. The first symptoms of the disease seem in a lot of hogs is a drooping appearance of the animals; they refuse to eat; the hair looks dry and has a dirty appearance; they have a hoarse cough; the bowels are sometimes costive, but generally the animal is affected with a diarrhea, perhaps always toward the last stages of the disease. The duration of the disease, as near as I can judge, is from five to fifteen days. In cases which I have dissected I find the lungs, liver, spleen, and bowels all more or less diseased. Some cases bleed at the nose; some go blind; some swell in the legs and break out in sores. A few of the latter get well; but none of those that bleed at the nose or go blind ever recover. There seems to be fever in every case.

I have also seen a great many hogs have a chill, as if affected with the ague. This disease is very fatal; indeed it is certain death.

Mr. JOHN M. LILEY, Taylorsville, Spencer County, Kentucky, says:

We have been visited in this and the adjoining county of Nelson by a disease among swine called "hog-cholera." It commenced a year ago this fall, and continued up to August last. During that time about two-thirds of the hogs and almost all the pigs died, so that there are only two or three small lots left for sale in this neighborhood. Those that I observed seemed to be attacked with inflammation of the lungs, accompanied with fever, which, if not resulting in death in a few days, continued as a slow pulmonary disease, with cough and very poor appetite, until the patient dwindled away to skin and bone, when death would ensue. Most of them died in this way on my farm.

I think the bowels of the animal are affected in very few cases of late years, and, therefore, the symptoms do not answer to those of cholera. We have tried a great many remedies—some patented, others vouched for by honest and sanguine men. None of them proved of any avail, however, either as a remedy or as a preventive.

As our swine had been free from disease three years previous to this fatal visitation, we had great expectations from them, and had increased the number. But they are about all gone; perhaps enough are left to supply the farmers with their own meat. We would be exceedingly thankful if some remedy or preventive could be discovered by which the disease could be controlled.

Mr. W. W. BARNES, Howard, Howard County, Indiana, says:

If there is any disease prevailing among farm animals in this county, except among hogs, it has not come to my knowledge. The so-called hog-cholera has, for the last year, prevailed to an alarming extent. In some cases the losses have been so great where large herds were held as to cause financial ruin. At this time a general feeling prevails against risking capital in this important staple.

The term cholera is generally used to designate the disease, but I doubt if a case of genuine cholera has occurred. In some localities a disease known as quinsy has prevailed—swelling of throat and jaws, attended with high inflammation. No remedy is known. In some cases, the knife was used in laying open the parts affected; but the recoveries were not as high as 10 per cent.

Pneumonia, or congestion of the lungs, is, I think, the real disease. After the hogs lose their appetites and refuse to eat they live from twelve to forty-eight hours. Death, when it comes, is instantaneous. The animals fall dead in the paths in which they travel, or die in the beds in which they sleep. Where they fall in snow there is not a sign of a struggle. They are always found on their bellies, as though their walk had been instantaneously arrested. All remedies seem worthless.

Mr. W. T. PACE, Centre, Kentucky, says:

There is no disease among farm stock in this section of country except among hogs. The disease prevailing among this class of animals has been very destructive. There has never yet been a remedy found that seemed to do much good. Mandrake-roots and red-oak bark, boiled down to a strong decoction and given freely, is the best remedy that we have found. The hogs are attacked in different ways. Generally an eruption of small red pimples breaks out over the entire body, but are most prominent on the breast and belly; their breathing is accompanied by a wheezing sound; their bowels are inclined to be too active. At least 90 per cent. of those attacked in this way die. Other symptoms are manifested by thumping in the sides of the animal. The hog becomes stupid, and will refuse to eat or drink anything for several days. This is not a very fatal disease. There is still another phase of the disease, in which the bowels are constipated. In those examined after death the fæces matter is found in hard, round lumps, the size of walnuts. It seems impossible for them to have an operation of the bowels. They live but a few days, and seem to suffer a great deal. Epsom salts, cream-tartar, and castor-oil are the only remedies that have ever done any good. The mortality is about 80 per cent.

Mr. S. H. BIDDINGER, Westport, Decatur County, Indiana, says:

Inflammation of the kidneys is a common disease here among horses. The early symptoms of the disease are those of fever; the horse is nervous and frequently looks around at his sides; stands with his hinder legs wide apart; is unwilling to lie down; shrinks when the loins are pressed, where some degree of heat is felt; the urine is voided in small quantities; frequently it is highly colored and sometimes bloody. The treatment consists in bleeding freely; next an active purge should be administered and counter-irritation excited as near as possible to the seat of the disease. For this purpose the loins should be covered with a mustard poultice. The horse should be warmly clothed. No diuretic should be given, but after the first effects of the purging have ceased small doses of white hellebore with tartar-emetic may be given. The animal's legs should be bandaged and plenty of water offered him. His food should also be carefully examined.

I have had some experience with the disease known as hog-cholera, and regard it as either a congestion of the lungs or of the bowels. A *post-mortem* examination disclosed the fact that those that run off at the bowels show a diseased condition of the liver and bowels, while those that are not affected with diarrhea die much sooner than the former, and present a highly congested condition of the lungs. Remedies that have proved efficacious in my practice are such as sulphur and coal-oil or sulphur and linseed-oil given in small doses twice a day until relief is found. Two drachms of sulphur to one ounce of oil is the proportion I use.

Chicken-cholera is first observed by a moping or stupid condition of the fowl. *Post-mortem* examinations show an enlarged condition of the liver. I have relieved fowls affected with the disease with a strong butternut-bark ooze mixed with thin feed. Small doses of calomel also relieve them.

Mr. GEORGE VIRGIN, Little Indian, Cass County, Illinois, says:

There has been no prevailing disease here among farm animals for the past three years except the much talked of hog-cholera, which has killed about one-fourth of the hogs of the county. The first symptoms are a severe hacking cough, constipation of the bowels, and loss of appetite. The hair of the animal almost stands on end, a high fever is manifested, which is soon followed by mortification and death. So far no very effectual remedy has been discovered. A good preventive is found in charcoal and copperas mixed with a little sulphur, common salt, and saltpeter. One pound of calomel sprinkled over some wet oats and placed in troughs for about fifty hogs, followed on the second day with a large spoonful of turpentine for each animal, is the best remedy yet discovered here. The turpentine should be given in slop, and the hogs kept away from water a day or two before giving the medicine, in order to give them an appetite.

Mr. IND. SMITH, Wellsburg, Chemung County, New York, says:

I have not had a great deal of experience with the disease among cattle known as the "western fever." Cattle shipped from the West to the Buffalo yards, in apparent good health when they started, have died of this disease soon after their arrival at the above-named point. I am of the opinion that in such cases the disease was contracted at the pens along the lines of the railroads. A neighbor who recently bought a car-load of these western cattle has already lost four by this disease. No doubt the cattle were in good health when they left home.

A good many stock-hogs purchased at Buffalo have died of some disease, perhaps of cholera. One gentleman has lost one hundred out of a herd of 225 head, while others with smaller herds have lost in about the same proportion. If these stock-yards were changed or thoroughly cleaned and disinfected, the ravages of the various diseases to which farm animals are subject might be greatly lessened.

Mr. WILLIAM T. HOLT, Colorado Springs, El Paso County, Colorado, says:

As to diseases of domestic animals in this State, I reply briefly that cattle here are pretty uniformly healthy. Out of a herd of over four thousand, owned by myself, I have not lost half a dozen head from sickness in the past four years.

There is, however, a poisonous plant growing here, and fast extending over the best stock-grazing portions of the State, which kills annually a good many horses, and threatens to put an end to the breeding of horses here at no distant day, unless some efficient antidote be speedily found. Already there are large areas of what was a few years ago the best grazing portions of the State (in the counties of El Paso, Bent, and Elbert) where it is now unsafe to turn out a horse or mule at any season of the year, and almost sure death to the animal to do so in winter when the grasses are brown and dry and this poisonous weed brilliantly green in color and full of juice. It is known here among ranchmen as the "loco weed," so named, I think, because its first effect when eaten is to make the animal crazy. Thousands of dollars' worth of horses are ruined every year in this State from the effects of this poisonous plant. It has not, so far, killed many cattle, for the reason that owing to the vast numbers of this class of animals it is rarely that any one of the number gets sufficient to kill him, and being apparently less susceptible to its peculiar influence than horses. It has been observed that the more valuable a horse is, *i. e.*, the more highly organized, the less "loco" it takes to intoxicate and finally kill him. No antidote has yet been discovered, and if you can set on foot an investigation which will result in determining a sure cure for a "locoed" horse, you will confer a great benefit upon the stock-growing interests of this community. While this weed has not yet spread abundantly enough to kill many cattle, it is believed to be only a question of time when it will do so, if not checked. The use of horses being indispensable to cattle-raising on the plains, this noxious weed indirectly imposes a heavy loss upon the cattle-grower. It also affects our rams, sometimes killing them outright, but oftener rendering them emaciated, crazy, and useless, but this far less frequently than in the case of horses.

The only diseases to which sheep are liable here are scab and "sore mouth," this last, so far as I know, affecting only lambs before weaning. Out of a flock of ten thousand sheep, owned by myself, these are the only diseases I have had to contend with, and I have found both easily curable. The cause of the sore mouth is not known here. It is not a general but rather a local and rare disease, and never fatal if properly treated. I have never seen it until this summer, when some 1,200 of my lambs had it. The lips are first covered with "chaps," followed by pustules which grow thick scabs. These extend gradually over the thin skin about the mouth and into the nose, making the face extremely sore and feverish, and prevent the lambs from nursing well or grazing. I had the pustules and the scabs scraped off clean and a solution of carbolic acid applied with a brush, which effectually cured it.

Mr. A. H. McCoy, Gentryville, Spencer County, Indiana, says:

In answer to your inquiries I shall only notice the diseases affecting hogs. I have been a breeder of hogs for forty years, and during that time have never known any disease among swine so fatal as cholera. This county loses from ten to twenty thousand dollars per annum by the disease. As I have been a breeder of fine pigs for more than twenty years, I have been unusually interested in the diseases of swine, and have been able to guard against every other disease but cholera. Mange is generally engendered by filthy quarters; thumps by general debility, mostly for lack of healthy feed; but cholera, beyond reasonable doubt, is a contagion, and is carried from herd to herd by hogs affected with the disease. Near twenty-five years ago, when the cholera first made its appearance in our county, I discovered it was nearing my neighborhood, and as it was very fatal, and fearing it was contagious, I fenced about six acres in on the inside of my farm, some eighty or one hundred yards from any outside fencing. The result was, I did not lose any of my thirty-five head, though my nearest neighbors lost from one-half to about all their hogs. Since that time my observation and experience have been the same.

Last winter I lost over twenty head of fine Berkshire hogs and pigs, caused by a gang of hogs affected with cholera being driven into my immediate neighborhood for the purpose of feeding on the mast, which was abundant. The symptoms, &c., are as follows:

1. A cough which lasts two or three days, and a strolling, restless disposition.
2. Vomiting, which generally lasts about a day; hog very sick.
3. Purging, generally, but not invariably, lasts two or three days.
4. After the purging ceases, if the hog is likely to recover, it will generally eat a little; but those that ultimately die seldom eat anything after the vomiting sets in. Those that die usually do so within from forty-eight hours to ten days.
5. After vomiting sets in the hog has a high inward fever, accompanied with chilly sensations, a symptom I discovered by observation. Snow was on the ground last winter, and it was very cold at the time my hogs were dying with the cholera. Very often they would leave their beds for the purpose of eating snow, which they would continue for a long time, though they had plenty of water; then they would pile together and shiver, which they will do even in warm weather if they have the cholera.

As to remedies, I have tried a number of the most popular ones without any favorable results; indeed, I am satisfied there is no cure. The best preventive beyond all doubt is the fencing-in system—*let no hogs run at large*. The next is the scattering system—have but few together. Farmers lose on an average about one-half the number of their hogs whenever the disease gets into a large herd.

Mr. J. Zimmerman, Mount Carmel, Wabash County, Illinois, says:

No diseases among farm animals have recently come under my observation, except diseases among swine. With the various forms of so-called hog-cholera I have had considerable experience in my own stock, and observation among that of my neighbors. The report of Dr. Detmers to the Missouri State Board of Agriculture, a year or two ago, contains, in the main, a better description of the disease than I could give, as well as the best remedial and preventive prescriptions I have yet tried. His statement, however, that hogs kept in small numbers, as by people in towns, are comparatively free from disease, is not at all borne out by the facts in this vicinity.

The greatest fatality is among pigs; but I am well convinced this is to a very considerable extent from mange, although denominated "cholera," with all other diseases to which the hog is subject. While induced in many instances by perfectly obvious causes, I think the mange in many cases is inherited, or is the result of injudicious breeding. For instance, I have one sow, now suckling her third litter, whose pigs in each case have been mangy, although treated as other pigs that remained free from mange. She has in each case been bred to her own sire; none of my other sows have been bred to a related male. It sometimes happens with me that a sow couples with a young, immature male; the progeny in nearly every such case are diseased.

The nesting of swine under barn-floors and the like, i. e., under any low, tight covering, where there is not free circulation of air about the animals, is, in my experience, a certain inducing cause of cholera.

I have had better success from the use of Dr. Detmer's remedies, namely, tartar-emetic and calomel (particularly tartar-emetic), and seclusion of the animal, than from any other. I have administered it to quite a number, and have called the attention of my neighbors to it, and know of no instance in which it has been administered that it has not been attended with beneficial results. I can hardly think of anything that has not been recommended as a cure for cholera. I have tried dozens of so-called remedies, sometimes with apparent success, but ninety-nine out of one hundred of these, I am positive, are called remedies on no sufficient basis of extended experiment. It may be so also with the above. So far my experience and observation are largely favorable to its efficacy.

I feed in a large wood lot, where there is plenty of water and shelter from cold winds. I throw corn on the ground by wagon-loads for the animals to run to when they wish, but never two loads consecutively at the same place. I break up the nests occasionally and compel a change of sleeping quarters. I feed, at least once a week, a mixture of salt and wood-ashes. I breed only from mature animals, preferring Berk-shires for mothers and Polands for sires, but lay particular stress on maturity of breed-ing stock. Whenever I find an animal refusing its food, or wheezing painfully, or with an appearance of thumping in its sides when it breathes, or nestling down and shivering as if it had a chill, I remove it from the lot as quickly as possible and feed it from two to four grains of tartar-emetic in a small quantity of potato cooked with a little grease to tempt an appetite. Whatever, if any, of these measures may be the cause, my swine have been measurably free from cholera during the past four years. Still, I recognize the danger that it may break out among them in a week, and also the paradox that if it were not for the losses by cholera there would be no profit in hogs.

I cannot give a reasonable guess at the average duration of attacks, so wide is the variance. I think at least sixty per cent. of the cases prove fatal.

Mr. W. H. MALONE, Marion, McDowell County, North Carolina, says:

On inquiry I find the opinion prevails that fifty per cent. of the hogs in this county have died during the last year, and that the fatality has been about as great in several other counties of Western North Carolina. The disease is called "hog cholera," but very little is known of its causes; still less is known of any effectual cure for the disease. The symptoms are often not discovered until the hogs are found dead; frequently from three to five head are found dead together. Sometimes the hog shows an indisposition to eat, and generally dies within two or three days after the discovery of these symptoms. The people have many remedies, but all have proven unavailable.

A disease also known as cholera has been quite fatal among chickens for several years past. They die suddenly—are often found dead in great numbers in the morning. No remedy for the disease has been discovered in this locality.

Mr. J. R. HOLSTON, Anderson, Madison County, Indiana, says:

During the past eighteen months we have had a fearful epidemic among our hogs, called cholera. It has been very fatal, and last year carried off at least four-fifths of all the hogs of the county. Some think the losses were even greater than this, but to be on the safe side I put the figures at four-fifths. For ten years past the farmers of this county have been raising for market from 25,000 to 30,000 head of hogs, and dur-ing the last eighteen months they have lost by this disease in this class of animals alone, in actual cash value, from $300,000 to $400,000. These figures are large, but they are below the aggregate estimate of some of our stock-raisers. In the years 1875 and 1876 we had partial failures of the wheat crop; so during the two years, with these various causes, we have had a signal financial failure, and it will take at least four or five years, with such crops as we have this season, to catch up again.

The symptoms of this so-called hog cholera are varied and complex, so much so, in-deed, as to render it very difficult to arrive at any definite conclusion. The first symp-tom among young hogs or shoats is a cough, accompanied by a kind of heaving or thumping in their flanks. This continues for a few hours or a day or two, when the ani-mal dies. Some mope around, lie in the shade, and refuse to eat. Those affected in this way live anywhere from two hours up to three or four days. Some bleed at the nose, some are constipated, while others are laxative. The last-named symptom is rare, and hogs thus affected generally get well.

There are numbers of so-called remedies and preventives, but all have proved abor-tive. Soft-soap, calomel, black antimony, coal-oil, dog-fennel tea, sulphur, sulphate of iron, &c., have all been used, but without effect. No specific remedy or preventive will ever be found until the origin or cause of this most fatal epidemic is discovered. The farmers of Ohio, Kentucky, Illinois, Missouri, Iowa, Minnesota, and Michigan are suffering to as great an extent from the ravages of this disease as we are here in In-diana.

Mr. HENRY C. MILLER, Westport, Decatur County, Indiana, says:

Farmers in this locality sustain heavier losses from hog-cholera, so called, than from all other diseases to which farm animals are subject. When once fully developed, the disease baffles all skill and every remedy. Preventives afford about the only relief yet discovered. Ashes and salt, given once a week, is a good preventive; so are sulphur and turpentine, administered in milk or slop. The following is regarded as a remedy of some value: One pound of black antimony, one-half pound of sulphur, one-half pound of copperas, and one pound of black pepper. Pour hot water over one-half bushel of shelled corn or wheat, and stir in the ingredients; then add one peck of wheat bran and a little salt. Stir well and scatter along the paths of the hogs, or on any place convenient to the hogs, where the ground is hard.

The malady seems to be a lung disease. The hog breathes with a jerk, the breathing becoming shorter as death approaches. In cases where they purge, the animal lingers from seven days to two weeks, but with other symptoms they die generally within from two to four days. It is more fatal among shoats and pigs, but often kills hundreds of fatted hogs.

Chicken cholera is very destructive among fowls. Preventives are more effectual than remedies. Lime in their food and water-troughs, and sand and gravel within their reach, will greatly conduce to their health.

Prof. A. A. HOLCOMBE, D. V. S., lecturer on "Special Pathology" in the American Veterinary College, New York, says:

In reply to communication received from you last month I can only give the facts relating to contagious pleura-pneumonia as it exists in the State of New Jersey. It has prevailed to a greater or less extent in some parts of the State for a number of years past. That it is spreading is attested by recent outbreaks in localities where heretofore it has been unknown. In September, 1873, an outbreak of this disease occurred on a large dairy farm at North Branch, Somerset County, New Jersey. It was treated by a quack of Somerville (in the same county), and nearly every case died. I saw three of the cases, and they were undoubtedly genuine cases of the contagious pleura-pneumonia. In June of the next year (1874) I attended an outbreak on an adjoining farm. About forty cows were affected. I treated thirty-three, five of which died. I made post-mortem examination of three and found all the lesions and post-mortem appearances belonging to the above disease. The treatment given the cases was simply general and special stimulants. The small mortality in this outbreak can hardly be attributed to the treatment, but rather to the exhaustion of the infecting virus. Isolation was strongly urged, but could not be effected owing to the failure of the community to appreciate its contagiousness. The cause of the outbreak is unknown to me outside of the testimony of the owners of the affected cattle. In both instances they had bought strange cattle, one or more of which were coughing and apparently not thriving. Undoubtedly this was the manner of introducing the disease, yet it needs confirmation. During the summer just passed a very serious and fatal outbreak has prevailed in the adjoining county of Hunterdon, in the neighborhood of Clinton and Lebanon. Of its cause I know nothing. The disease is a terrible scourge to some localities of that State. An investigation of its cause and the best means of stamping it out is no doubt a subject worthy the attention of the Department of Agriculture.

Mr. Jo. ABBOTT, Hillsborough, Hill County, Texas, says:

1. I will say that my observation, which is supported by information I get from several well-informed gentlemen, is, that horses and cattle which run at large on our prairies are entirely free from disease of any kind.

2. That horses which are kept up for use are sometimes troubled with bots or colic. These cannot properly be said to be diseases; but instances of these complaints are rare. For the first, a drench made by dissolving about one-third of an ounce of bluestone in water sweetened is regarded as a specific. For the latter, one-half pound of bi-carbonate of soda, dissolved in water, is frequently used with good effects. In violent cases this is repeated once or twice.

As to milch cows and oxen, I can say I have known neither to be troubled with any kind of disease, and I have owned a number of each kind for several years past.

Hogs are frequently affected with cholera, which, at times, assumes the form of an epidemic among them. In seasons of this kind the loss is often 50 per cent.; but I will say that I have known of no cholera among hogs during the past twelve months. I am not informed of a remedy for this disease, although several experiments have been made.

Fowls, especially chickens and turkeys, are frequently visited with cholera. I have never known a fowl to be cured after the disease was fully developed, though many trials have been made. My observation is, that if fowls are fed on onions, mixed with other food, or if you can induce them to feed on the onion while growing, as they sometimes do, they will never have the disease. I believe the onion to be a sure preventive.

Mr. R. M. Mumford, Princeton, Gibson County, Indiana, says:

The farmers of Gibson County have lost thousands of dollars this year by what is termed hog-cholera. The first symptom of the disease is a cough, then the animal becomes stupid, refuses to eat, and generally dies within from three to six days from the appearance of the first symptoms. Corn soaked in lime and fed to hogs is said to be a preventive. Copperas, sulphur, and ashes are also said to be of some value as preventives. No cure has as yet been found.

If anything can be done by way of investigation by your department that will afford relief, it will be thankfully received by the farmers of this section of the country.

Mr. C. W. Johnston, Chapel Hill, Orange County, North Carolina, says:

Distemper has prevailed to some extent among horses and mules in this locality. The duration of the disease averages about one month. Not one in one hundred of the animals attacked die of it. An efficient remedy is found in the inhalation of smoke from burned tar and feathers. Asafetida used on the bit and in the trough will be found a good preventive.

Murrain has prevailed to a small extent among cattle, with fatal results. There seems to be no remedy for this disease.

Among hogs the cholera has prevailed to an alarming extent. Sulphur, turpentine, copperas, &c., have been used as remedies, but none of them have proved efficacious.

Cholera has also been very destructive among fowls, and, as with hogs, all remedies seem to be ineffectual.

Mr. H. A. Cutting, Lunenburg, Essex County, Vermont, says:

I would say that the use of powdered lobelia, or ipecac, in all cases of epizootic or colds in the heads or throats of horses have, in this section, been beneficial. The manner of use has been to take a large spoon and put into it a drachm of ipecac or two drachms of lobelia, and, after opening the horse's mouth and drawing out his tongue, scatter the powder as low down on the roots of the tongue as possible. In this way it is mostly swallowed, and yet some is worked about the mouth and throat, causing an increased secretion and almost immediate benefit.

Hogs have died to some extent this summer—perhaps one-sixth of all. The disease seemed induced by constipation, and after the cause was discovered all were saved by giving them common salt. Those not sick were given salted food, and all went well.

Mr. George C. Eisenmayer, Mascoutah, Saint Clair County, Illinois, says:

We have no general diseases among farm stock in this county, except cholera among hogs, for which there is no known remedy. There are also occasional cases of cholera among all kinds of domestic fowls, for which no remedy has been found.

Mr. B. Whitaker, Warsaw, Hancock County, Illinois, says:

It is with much gratification that I learn that the diseases affecting farm animals is about to receive attention. The losses in this county from hog-cholera alone are estimated in cash value at $30,000 per annum. In a recent report of the State Board of Agriculture the disease was said to exist in eighty-eight counties of the State, and from authentic and well-digested reports the annual loss was estimated at $7,880,060. The terrible fatality of this disease and the great losses sustained thereby is the strongest argument that could be offered in favor of a speedy investigation into its causes. Remedies without number have been prescribed, but without any appreciable effect. The disease, in its various forms, is veiled in so much mystery that a correct diagnosis is rendered very difficult. The symptoms generally, as I have observed them, are about as follows: First, the hog becomes stupid and refuses to eat, sleeps a great deal, and dies within a few days. Second, it may be constipated or exactly the reverse. Where diarrhea prevails the hog may die soon or it may linger along for several days, all the time losing and shrinking in flesh. Sometimes animals affected in this way recover, but they remain poor, gaunt, and apparently shriveled up. Young hogs are generally affected with a hacking cough and a noticeable jerking pulsation in the flanks at every inspiration of breath. Pigs and shoats will sometimes linger for weeks with these symptoms. Still another symptom is observed in cases where the hog seeks seclusion, with every appearance of a severe cold or chill. It will crouch into the smallest possible compass, apparently for the purpose of securing warmth. Some hogs are attacked with vomiting and purging, which symptoms continue until death ensues. The disease is more fatal with fatted hogs than with any other class. Many of these drop dead without a struggle, and without any visible symptoms of disease.

Intestinal worms may possibly have some connection with the diseases which affect swine. I was informed by a gentleman who performed the operation, that in spaying some hogs last year he found the intestines of one greatly distended with worms. He opened them and took out fourteen long, large worms, and closed the opening without completing the operation of spaying. The hog lived and did well. Another case, where the intestines were opened, a large number of worms taken therefrom and the hog afterward spayed, the operation proved fatal. Proof is abundant that intestinal worms are common to most hogs, both in sickness and in health.

Diseases of fowls exist in almost every community and locality. It has not been so prevalent this season as in past years. It is, perhaps, contagious, as healthy fowls brought from other places and allowed to run with diseased ones are soon infected. Guinea fowls, ducks, and geese are exempt from the disease so far as my knowledge extends.

I neglected to state in the proper connection that all hogs affected with any of the above symptoms refuse to eat, hence the difficulty of administering medicine.

Mr. W. M. GREEN, Jamestown, Russell County, Kentucky, says:

There are many complaints of cholera among hogs, but I am seventy years old and have never had a case in my herds. I have sometimes had hogs affected with quinsy— a swelling of the throat. This disease is generally fatal. Lice no doubt cause many diseases which ultimately prove fatal. I have had a good many sucking pigs and small shoats die of a disease resembling consumption. The first symptoms are those of wheezing and coughing. They then become constipated, refuse to eat, seem very stupid, take the thumps, and soon die. My grown hogs are generally healthy. I feed from seventy-five to one hundred and fifty every year, and scarcely ever lose one. I frequently give them copperas, sulphur, and soda, about one-half pound of each to every twenty hogs. This is mixed with soft-soap and rye-meal. For quinsy I give spirits of turpentine, or common tar mixed with meal.

A disease called cholera is very fatal to chickens in this locality. I never had it in my brood until this fall. They have died very rapidly; indeed, it seems they will all die, as we have no remedy. When attacked with the disease the fowls become stupid, refuse to eat, run off at the bowels, and soon die.

Mr. W. J. MOORE, Larkinsburg, Clay County, Illinois, says:

I am happy to report that there is no prevailing disease among any of the domestic animals of this part of the country; all are in a healthy condition at present. Cholera prevailed to some extent during the early part of the past summer among hogs, but it did not assume an epidemic form and soon abated. Its abatement was not attributable to any specific treatment.

Mr. S. V. PICKENS, Hendersonville, Henderson County, North Carolina, says:

In this locality, where the atmosphere is mountainous and the water pure, the most of the ills to which horses are liable are, either directly or indirectly, the result of mistreatment, except, however, the epizootic and other distempers, not very prevalent at any time in this section. Among the most common diseases here are the gravel, scours, glanders, and colic.

In case of gravel the horse manifests great pain; when standing will stretch his legs far apart; when lying the animal rolls much upon his back. When thus affected the horse must be relieved in a few hours, or death will ensue. As a remedy, take two eggs, pour out the yellow through a small hole broken in the shell, then fill the shell with spirits of turpentine, and make the horse swallow the whole. Some inject onion juice up the water-organ with good results.

Scours are generally caused by excessive exercise or over-feeding with green food. This causes over-heating, which is followed by loose discharges from the bowels, producing general debility accompanied with great suffering. A dose of spirits of turpentine or tar ooze will generally relieve the animal by checking excessive discharges, after which drench freely with warm sage or pennyroyal teas.

Glanders affects the roof of the horse's mouth, produces great soreness, and renders it very difficult for him to masticate his food. Sometimes some of the bars in the roof of the mouth become a gristle. Bleeding in the roof of the mouth and frequent swabbing with a strong solution of copperas and alum is our remedy.

Colic may be caused by excessive work, irregular and excessive eating, drinking, &c. It is indicated by the strongest manifestations of pain, great restlessness, continual walking, rolling or pawing, and body swollen. The most speedy cure known to us is to "rake" the animal and bleed in the neck and mouth. Then give him freely of warm teas by drenching, with soda dissolved in it. This disease does its work usually in a few hours.

We believe most horses have bots in them, but that their ravages are seldom committed upon an animal when in good health. Therefore, when a horse is debilitated and his whole organization deranged by disease, is when the bots begin their work. This is known by the great restlessness of the horse, and the resting of his nose upon his flank. One-half pint each of whisky, lye, sweet milk and molasses well mixed and poured down the horse in time, is almost a sure cure, but should be followed in one hour by one-half pound of salts, to be repeated if ineffectual. These remarks have special reference to this immediate locality, but are alike applicable to the mountainous region of western North Carolina.

Before closing my statement relating to horses, let me advise the free use of salt and lime, or wood ashes, mixed in food. It serves as a preventive for many of the diseases common among domestic animals of this section.

Our cattle seldom die of disease, save the "hollow-horn," more justly called "hollow belly," since the latter is generally the cause of the former; and distemper, believed to be contagious and almost invariably accompanied by what we term the "distemper tick," great numbers of which get upon the cattle about the time and in localities where the disease rages. It is thought to be communicated by grazing where affected cattle have lain or grazed. It is also said that a cow may have it in its system and communicate it to others and show no symptoms in themselves. The free use of sulphur internally and kerosene oil externally serves as a good preventive, in which alone is safety.

Hogs are sometime affected with cholera, which is supposed to be transmitted from one to another. So very fatal is this disease that perhaps 80 per cent. of the hogs attacked with it die. Tar and copperas are good preventives, used in food. Kerosene oil and blue stone are as good remedies as we know of here.

Dr. JOHN KENNEDY, Paragon, Morgan County, Indiana, says:

Hogs being our staple production, I shall treat of the various diseases affecting this animal, all of which are called cholera. In my opinion there are three distinct diseases, viz: Lung fever, (pneumonia,) erysipelas, which may affect any one organ or the entire organization of the animal, and enteritis or enteric fever, a disease similar to hospital or camp or typhus fever in the human system.

The former is mainly brought on by exposure to changes of weather. The two latter are epizootic and contagious, and so closely resemble each other that I shall not attempt a distinction, as they are quite generally considered the same disease. I shall simply give distinctive symptoms sufficient to enable the ordinary farmer to know what ails his hogs.

In the colder seasons of the year, when the hogs are inclined to pile up to sleep (not being protected, as is nearly always the case in our vicinity), it is noticed that some of them do not readily come up for their morning feed, and when they are driven up they seem stupid and not inclined to eat. They may have a cough, or this symptom may not show itself for a few days further along. They are thirsty from the beginning and the cough, which appears sooner or later, may be accompanied with bleeding at the nose and mouth, which is an evidence that the lungs are seriously affected. When this latter symptom appears it may be taken as an evidence that the animal will soon be ready for the dead-hog man. The symptoms invariably indicate lung fever. The best treatment is to at once separate the well from the sick ones and if possible provide shelter and protection for all. If you have too many in the herd take out those that are positively healthy and put them on the market, and thereby reduce the number until you can afford shelter and protection for the remainder. A cheap shelter and protection may be constructed by boarding solid your fence so as to shield them from the chilling effects of the northwest winds. Make a cover slanting from the top inward, and throw in stalks and husks for bedding. Further on I shall give a diagram for a barn, such as every hog raiser should have.

The next thing in the general treatment of the disease is to cease feeding everything except slop made from corn meal, with sufficient salt added to make it palatable. They should not have water oftener than three times a day. I would give from one to two pints of water from pine tar, adding five to ten grains of nitrate of potash to the pint. During the active stages of the disease and in convalescence, which will take place within from five to seven days, I would use chlorate instead of nitrate of potash. With this simple treatment more hogs will be cured than in any other way that I have known tried. If thought proper, however, a small amount of copperas may be given during convalescence, say from two to five grains to the hog three times daily in their swill or slop. As a preventive for those not affected nothing is better than the tar water mixed with chlorate of potash. As a disinfectant copperas water, or charcoal and wood ashes, may be used. Carbolic acid, if not considered too costly, may also be used in the proportion of one-half ounce to a quart of water. With this the beds should be sprinkled two or three times a week, using a common sprinkler or a wisp of straw.

In cases of erysipelas the hog will appear indisposed and rather mopy. At first its bowels are somewhat constipated and its fæces dry and hard. Within a few days diarrhœa, though not always a symptom, may be noticed; red or bluish-red spots will appear on the skin; swelling will set in, and, if the hog does not soon die, the hair will begin to fall off, and the skin, in some cases, will become surfeited and even crack open. It will thus linger along for thirty or forty days, and sometimes recover after it has been given over to die. This disease is liable to affect the vital organs, and when it does it runs a rapid course, proving fatal in a few days or resulting favorably in a comparatively short time. The distinguishing symptoms in this and enteric fever or inflammation of the bowels are, instead of the red spots on the skin, an eruption of red specks appear, and vomiting and diarrhea are generally present within a very few days after the attack. If not properly treated it is equally fatal with the others. As in other cases, I would advise separation of the sick from the well ones, and in cold weather, shelter and protection, also observing like rules as to feed and water, using tar-water with carbolic acid. One ounce of the latter to a barrel of water, and one gallon of the water to each hog per day in addition to three quarts of thin corn-meal gruel to each hog, will be found the best treatment. For those that have diarrhea, one-half teaspoonful of muriate tincture of iron may be given three times a day. A small amount of carbolic acid for the well ones may also be given. I cannot give the proportion of hogs cured by the above course of treatment, but so far as tried it has proved very effectual. To be healthy, hogs should have a fair degree of cleanliness, and where they do not have access to running water, the pools where they wallow should be disinfected once a week by the application of either lime, wood-ashes, or copperas.

Herewith I give a diagram of a barn owned by Mr. Jesse Lockhart, of Niantic, Ill., which he erected for the protection of his hogs. Two years ago this gentleman informed me that he had been using this barn for three years, and that during that time, notwithstanding he had handled several thousand hogs yearly, he had not lost one from the so-called hog-cholera.

The foregoing design comprises two cribs with a drive-way and scales between, making a main building forty feet square and fourteen feet high, with gables at each

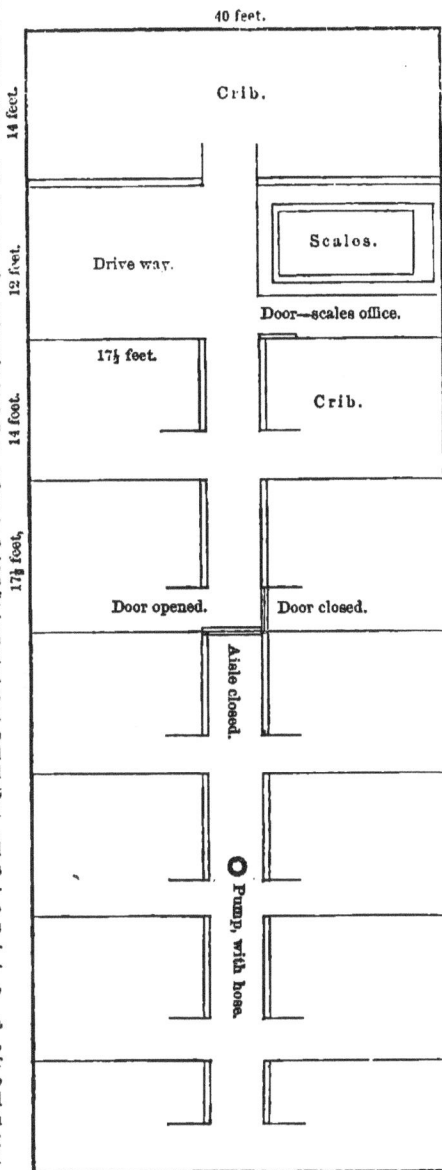

end of the drive-way. The pens attached and contiguous thereto are covered at right angles with the cribs; these pens—six side by side, or twelve in all—extend the building one hundred and five feet, which, added to the other apartments, makes the entire building cover a space of 145x40 feet. The pens, which are about six feet in height, have windows to each, with shutters, and may be closed tight or ventilated at will. The inner walls of the pens are four feet high, and the aisle and doors five feet wide. The doors open in opposite directions, and when one is opened it closes the aisle, so that hogs can be changed from one pen to another by simply opening two doors. Each pen is provided with a trough, and near the center is a force-pump, supplied with a rubber hose long enough to reach to any part of the pens. With this apparatus, Mr. Lockhart informed me, he thoroughly cleaned his pens once a week. The cobs are scooped up and taken out daily, with all other refuse matter, and dumped out to the stock hogs, which are fed in adjoining lots, each lot containing fourteen acres. One of these lots is planted to soft maples and the other to black-walnut trees, the trees now being about seven years old.

Mr. LINFORD H. HAWES, Woodlawn, Jefferson County, Illinois, says:

We have no diseases of an epidemic nature among our farm-animals other than cholera among hogs and chickens, the diagnosis of which is not different from that heretofore published by the commissioners of the State of Missouri. There is no specific remedy in use among our farmers, though bicarb. soda has been used and is claimed to be such remedy.

As a preventive for the disease among fowls, brimstone, *i. e.*, roll-sulphur, has been placed in their drinking-vessels to impregnate the water. Bicarbonate of soda is also used as a remedy, and, it is claimed, with excellent results. However, I believe it is generally admitted that fowls are exempt from diseases of all kinds if kept free from vermin. When on a new farm or cleared land they have access to a plentiful supply of insects and grubs found in decayed logs and brushwood, which argues that a liberal allowance of fresh meat, together with plenty of coarse gravel and scrupulous cleanliness, is all that is necessary to insure exemption from disease.

We have in some localities " milk sickness," with which domestic animals are liable to be attacked, and from the use of beef, milk, or butter, the disease is imparted to mankind. It is useless for me to repeat the symptoms so frequently described heretofore. The cause is as much in doubt now as it was at the first settlement of the country. However, it is claimed by a few who have given the matter consideration that the disease has its origin in poison by cobalt or black oxide of arsenic. The exhalations from the soil containing the poison gather upon the herbage or impregnate the water, and thus are transmitted to whatever partakes of either. Proof is offered in the fact that acid sulph. aromat. is an antidote for poison by cobalt, and it has been used with good results in what seemed hopeless cases of this poison among oxen.

Mr. HENRY GRUBE, Beaver Creek, Bond County, Illinois, says:

There is no general disease among any class of farm-stock except among hogs and chickens. Every boy knows a cholera hog or a cholera chicken when he sees it, but the most shrewd and knowing differ widely as to the cause of the disease. From my own experience I am satisfied that, with proper care and such means as are within the reach of every hog-raiser, no one need have cholera among his hogs. Through carelessness I have lost a number of hogs, which has only occurred with me in a busy time. My plan is to prevent, which can be done by placing common wood-ashes in a trough or on the ground, with salt scattered over them, and some kind of grain, bran, or meal thrown on top of that, say once a week. Besides this, place some stone-coal within reach of the hogs, and my word for it you will have no cholera. Burned bones will also be found a good addition. Arsenic is curative, but "one ounce of prevention is worth a pound of cure."

It is seldom we are afflicted with chicken-cholera, and therefore I have given the subject little or no attention.

There is no limit to the duration of these diseases. The average fatality is about 95 per cent.

J. BRICE, veterinary surgeon, Eric, Pa., gives the following diagnosis of a fatal cattle-disease which recently prevailed in that locality:

In reply to your inquiry respecting the cattle-disease which prevailed here for a short time, I would say that, so far as we know at the present time, it has completely subsided. Nearly all of the animals attacked died of the disease in length of time varying from a few hours to not exceeding five days. In some cases so rapid was the disease that animals thought to be in perfect health in the evening were found dead in the morning. (These sudden deaths were known only by hearsay.) The animals attacked, so far as known, were all milch cows, and the only ones that recovered were young cows. Although some few others recovered, it is believed they were not suffering from the specific disease, but some disease consequent on overfeeding, and in some cases from lung disease. There may be a cause for all the animals attacked being milch cows, as the disease was confined to the city altogether, and few other cattle are kept in the city. In a barn in which was the greatest fatality there was a bull which stood through it all, his companions dying to the number of seventeen. Only two cows were left, one of which did not have the disease. The other, a young cow, recovered from a slight attack.

The disease was certainly splenic fever, charbon, or anthrax. The symptoms were extreme restlessness, loss of appetite, but not complete; thirstiness; faeces natural at first but frequently diarrhea afterward; the urine profuse, and during the latter part of the disease dark red or bloody-looking. The animal gave evidence of intense internal pain by her arched back, hanging head, and, if at liberty, by her constant moving, or, if tied, by pushing her nose into a corner and breathing laboriously. Although at first the animal had perfect control of her limbs, they became first weak, then staggering, and finally lost their power completely. She would then fall down, and, after a few ineffectual attempts to rise, would lie helplessly moaning until death relieved her suffering.

As to treatment, everything that was tried availed nothing. The fever steadily progressed to the end. Further research would seem to be something most devoutly to be wished for, and we hope that some measure of success may attend every attempt to find a cause and a cure for so fatal a malady.

All the cases in this section have been in that particular portion of the city where the cattle pasture over a large common, in direct communication with the railroad and cattle-yards, and where a number of Texan cattle were grazing after removal from the cars during the period of the recent railroad strike. Soon after that time the first cases were noticed, but the cool weather early in the fall appeared to check the disease, only, however, to break out with greater virulence during the hot weather in the latter part of September.

Mr. S. P. THACKER, Vienna, Johnson County, Illinois, says:

Horses and mules have been affected with what is known here as periodic opthalmia. The first cases that came to my knowledge were in January, 1875. Only two or three cases occurred then. The disease has since become prevalent, so that there are numerous cases now within my knowledge.

The animal is attacked with inflammation and swelling of the eyes, nearly invariably beginning in the left eye; then, within from twenty-four to thirty-six hours the right eye is attacked in the same manner. The eye runs a clear, thin, watery fluid, and in some cases matters. While inflammation is in the eye the light seems to be painful to that organ. The inflammation lasts three or four days; then subsides, leaving the pupil of the eye of a milky color. In the course of four or five days the eye becomes apparently well again. The animal becomes nearly blind during the attack, but can see again very well after the attack is over. Some have become blind in one or both eyes after the fourth or fifth attack, which occurs at intervals of from three to seven weeks. The cases of longest standing seem to become more severe and of longer duration; but the attacks are not so frequent.

It is thought by some of our veterinary surgeons that the disease is hereditary, but I notice that stock of entirely different pedigrees are attacked by it. Bathing the eyes in warm salt-water appears to be of more advantage than any other remedy yet tried. This allays the inflammation, but does not prevent the recurrence of the disease.

Mr. W. J. BANKS, Elizabethtown, Hardin County, Illinois, says:

There is no disease affecting farm animals in this county except a disease among cattle mentioned in a former report to your department; but I am still unable to learn the name or nature of this disease. Cholera is prevailing among swine to some extent, but we are never *entirely* rid of this scourge in this county.

Mr. J. H. OAKWOOD, Catlin, Vermillion County, Illinois, says:

Hogs are very unhealthy here and are suffering from a disease called cholera. This disease manifests itself in various forms. Sometimes it seems to be congestion of the lungs, at other times sore throat, at another time rheumatism, and still again an affection of intestinal worms. Sometimes the disease takes the form of chills and fever, and then the hogs will lie in heaps in the warmest weather as they do in the winter season for the purpose of keeping warm. These diseases are all designated as "hog-cholera," and no remedy is at present known. It is generally conceded that to drive the sick hogs rapidly and heat the blood, and give but little food, is about as good a remedy as any other. In some forms of the disease tartar-emetic has been used successfully. Various other remedies are used, but all fail in a greater or less degree.

The animal usually lives but a few days after being attacked. In some cases hogs become affected and lose flesh; the appetite appears good, but it seems impossible to fatten them. In this condition they sometimes live for months. In cases like these the better plan is to turn on grass and give no food, and in a few months the animals may again become healthy. Seventy-five per cent. of all the hogs attacked by this disease die, and full ten per cent. of those that reach maturity in this county die of some disease other than the above.

Mr. SAMUEL PRESTON, Mount Carroll, Carroll County, Illinois, says:

Diseases among domestic fowl have been very fatal during the past and a few preceding years in this locality. My wife is of the opinion that a liberal mixture of wheat-bran with other food is a preventive of disease. It is also excellent for hogs confined chiefly to a corn diet, by keeping them from becoming constipated.

With the exception of distemper, which, in a few cases, has proved fatal, horses have been pretty free from contagious diseases. Since the epizootic passed over the country a few years since, a large fatality has befallen young colts. Probably fifty per cent. of these young animals have died the present season. Some attributed it to the effects of that disease.

A strange disease has attacked and proved fatal to my lambs during the past three seasons. It comes upon them about midsummer. From apparent health they die within from three to four hours. They are first noticed lying down in a natural position, separated from the rest of the flock. A fit or spasm seizing them, they will throw themselves upon their sides and, with eyes set, will soon expire. In 1875, I lost forty; in 1876, fourteen; and this season, four. None recover that are attacked. I have found that weaning the lambs early checks the disease.

Mr. W. O. Millard, who resides about two miles southwest of Milledgeville, and who is one of the largest and most careful stock-raisers in this locality, has been very unfortunate with his large stock of hogs during the past summer. He has lost one hundred and ninety head, sixty-one of which were large hogs, the remainder shoats. He claims that the disease which decimated his herd was nothing more nor less than typhoid fever, and thinks it will yet extend far more than it has in this and adjoining counties.

Mr. M. DAVENPORT, Oxford, Calhoun County, Alabama, says:

Cholera among hogs is the most dreaded and fatal disease we have to contend with here as affecting any class of farm animals. It is seventeen or eighteen years since it made its appearance in this locality, and it now passes over this country as an epidemic about every other year. I know of no remedy for it, neither can I give any information in regard to its cause. Some years ago I lost three hundred head of hogs by its ravages in the short space of fifteen days. The disease has prevailed among my hogs six or eight different times, doing great damage at every visitation. I have tried almost every prescription recommended as a remedy without any beneficial results whatever. If, in your proposed investigation, you succeed in finding either a preventive or a cure for this terrible malady, you will receive the thanks and blessings of hundreds of thousands of stock-raisers in this country.

Mr. J. ELLWOOD HANCOCK, Columbus, Burlington County, New Jersey, says:

I have had some experience with pleuro-pneumonia in cattle, having lost one-third of my herd from its ravages in 1861, when I succeeded in eradicating the disease after a duration of about six months. I had a second visitation of the malady in my herd in the early part of 1876, when I lost six head from a herd of twenty-three. My experience is that it runs its course in not over three weeks after the animal becomes so much affected as to prevent its eating—usually in a shorter time. Of the animals affected, I am satisfied not more than one-third will recover. I applied to a veterinary surgeon, who prescribed a powder which I think was a benefit, giving it, as I did, to the whole herd as soon as it was ascertained the disease was present. After the disease is fully developed in an animal I have very little faith in medicines, as a large proportion will die with the best treatment. Although my whole herd was not really sick, the larger part of it showed signs of the disease; some only for a few days, however. It remained among my cattle for about four months. I am of the opinion that on both occasions the disease was introduced by cattle purchased by me. The first case showed itself in about six weeks after the introduction into my herd of the infected animal; in the second case it was at least four months. I regard this as the worst feature of the disease—it remains dormant in the system of the animal for so long a time before it is imparted to others.

During the past few years this terrible disease has caused great loss to farmers in this section of the State. Many have had to contend with it, and numbers have suffered heavier losses than I have.

Farmers in this locality are also suffering great losses from chicken-cholera. The fowl is taken with diarrhea and sits moping about for a few days and dies. But few of those affected recover. Many preventives have been tried. I believe cayenne pepper, asafetida and composition powders, used freely in the feed, are useful as such. Cleanliness in roosts, gas-tar, carbolic acid, &c., are useful; but the preventive or remedy remains to be discovered which will give absolute security.

Mr. JAMES H. SWINDELLS, Lancaster, Dallas County, Texas, says :

We have not been troubled with diseases among any of the lower animals except among hogs and chickens, both of which were, and now are, affected with what is termed cholera. Until a year ago the hogs in this locality were not affected with cholera. The disease was brought here by the importation of stock from Wise, Montague, Parker, and Johnson Counties, a tier of counties lying in the lower Cross Timbers, west of this point. When they arrived they were herded with hogs raised here. In less than a week the imported hogs became diseased and commenced dying rapidly. The affected ones were separated from the others and various remedies were made use of to check the disease and, if possible, cure it. None of the remedies used seemed to be of any benefit, and nine-tenths of those affected died. The disease soon spread to the native stock, and since then (last fall) there has been more or less of the disease present.

The symptoms observed are as follows: Indisposition to move about or to eat; lying down most of the time;.diarrhea, with excrements first of a natural character, but gradually getting darker until the evacuations became almost black; fever, the temperature in some cases running up to 10½° F., but generally to about 102°. Before death the animal would vomit a dark-green or black fluid, swell up, and the odor emitted would be very offensive.

The only effective way of checking the disease would seem to be to separate the diseased animals and put them into a clean lot having running water in it. I had a few hogs which were taken sick with this diarrhea. In a day or two the discharges became of a light-green color, and very thin. I relieved all of them but one (I believe seven were attacked) by the administration of calomel. For a hog weighing one hundred pounds I would mix one dram of calomel with a handful of meal and a little milk, and let them have that much in the course of twenty-four hours. They would generally eat a little at a time until the whole is disposed of. The calomel did not seem to purge. On the contrary, the bowels would check up, and in from one to two days the animal would commence eating corn and would get well without any further trouble. The one which died would not eat the meal in which the calomel was mixed.

Mr. W. DUNLEY, Hennepin, Putnam County, Illinois, says:

A disease called cholera has prevailed to a great extent among hogs in this locality during the past few years. Many of our farmers have at different periods, and within a very short time, lost most of their stock by the ravages of the disease. No positive remedy has as yet been discovered.

During the past summer I lost about eighty hogs by the disease. I used all the different remedies recommended, but they continued to die daily until I was told that oats was a specific. I at once commenced feeding dry oats, and no more died. Three were sick when I commenced feeding the oats, but they recovered, and I have lost none since.

A great many fowls have died in our vicinity of a disease also called cholera. No remedy or sure preventive has been discovered for this malady.

Messrs. DANIEL A. and JACOB MILLER, Farmington, Davis County, Utah Territory, write as follows:

Sheep are the principal stock product of this locality. Among this class of animals a disease prevails called scab, for which the following remedy is used: One peck of unslaked lime and twenty-five pounds of sulphur dissolved in water. A tank or hogshead is filled with the water in which these ingredients have been dissolved, into which the sheep are dipped. These dippings are generally required once or twice a year. Another remedy is to make a solution by adding to water sufficient for the purpose one pound of tobacco, one-fourth pound of gunpowder, and two ounces of arsenic.

This solution is poured on the backs and other affected parts of the sheep. Sometimes a small amount of red precipitate is used, but this is considered dangerous.

But few hogs are raised here, and they are generally healthy. However, there have been some cases of cholera, but how the animals were treated I cannot say. Fowls are generally healthy, with no prevailing disease.

Mr. T. H. BARR, Augusta, Macon County, Illinois, says:

Hogs have been destroyed every year for the last twelve years in this locality by a disease known as "hog cholera." The disease has never, as far as I have been able to learn, prevailed in the open prairie without our being able to trace it to some marked source of contagion, such, for instance, as native swine coming in contact with hogs brought in from localities where the disease was prevailing. The disease prevails almost continuously along the timber belts on the water-courses, owing doubtless to the fact that hogs are suffered to run at large, while many careless persons throw the dead carcasses of the animals into the streams, thereby spreading the disease along the whole length of the water-course below.

Where the disease breaks out spontaneously as it were, the symptoms are a violent cough attended with high fever. I have been told that on examination of such cases after death the lungs were found in a decayed or rotten condition, while the other vital organs presented little or no derangement. Such cases originate in close, ill-ventilated quarters, such as are found under the floors of old buildings or about or under straw-stacks. The carcasses of such, if eaten by well hogs, or even the droppings from them will communicate the disease in a more intensified form and fatal character than that described above. With the latter cases the hogs die more suddenly than in the first instance, sometimes within twelve hours from the attack, while the former will often linger for days. In some cases the latter, in addition to the cough and high fever, will be extremely costive; in other cases the animal will be affected with an active diarrhea. Some will swell up about the ears, the skin will crack open and the blood will ooze therefrom. All or nearly all of those thus affected die. The few that do recover had better die, as they rarely become thrifty again.

We have never yet found a remedy that will effect a cure. The best informed stock-raisers are of the opinion that relief must come, if it ever does come, through preventives rather than through remedies.

Those of us who have been most successful in keeping our hogs free from disease have done so by giving them good, comfortable, clean, well-ventilated quarters, and as a general thing those who most nearly meet these conditions have the best success.

Fowls are affected and thousands die annually by a disease known by the name of cholera. The symptoms are about as follows: Two or three days before death they will appear droopy and stupid; eat but little if any; become very thirsty; have a very active bowel-complaint, and finally drop down dead. Another symptom is seen in the gills and comb of the fowl, which become pale soon after attack. The only remedy that has yet been employed with success here is to rid the premises of fowls for twelve months. After that they may be kept again for a few years free from disease. There are those who are of the opinion that fowls exhaust something on the premises that the system requires, and until that constituent is replaced they cannot live and thrive.

Mr. R. K. SLOSSON, Verona, Grundy County, Illinois, says:

The hog seems very much more subject to fatal diseases now than he did forty years ago. To arrive at a correct etiology of the diseases of this animal, which forty years ago were unknown, we are forced to notice the then physical condition of the animal as compared with his present, tracing the changes which have been effected by confinement, change of food, and the practical method of producing new varieties which shall take on the greatest number of pounds of muscle and fat in the shortest time. Of all the domestic animals the hog is the most easily made to undergo changes of form and temperament, and hence it is that the varieties of the hog are continually increasing. New breeds, well advertised and puffed, are multiplying, and the great and only object appears to be to find a variety that shall eclipse all others in maximum weight at the earliest possible period of their existence. In the insane pursuit of gold stamina of constitution are lost sight of, and the hog-raiser who has three hundred head to-day in four weeks' time may be reduced to half a dozen head. He sustains a loss of $3,000 from the emasculated system of the hog making them susceptible to disease which a healthy and strong constitution will not take on. A change of constitution was doubtless brought about in part from confinement, a condition unknown to the hog before domestication. Confinement, as all physiologists know, decreases muscular growth and strength, and the nervous energies are correspondingly weakened. On the heel of this a change of food takes place. Indian corn is fed in many parts of the country to the exclusion of those kinds of food upon which he had previously lived, for hundreds of years perhaps, and corn is almost exclusively fat-producing. This combination of new circumstances and conditions necessarily produces physiological changes in the system, and these changes being, to say the least, partially abnormal, the body is prepared to take on diseases which were originally unknown to the hog. It is these changes which create a predisposition to disease which hitherto inoperative causes have failed to develop, but now being brought into action the enervated system falls an easy prey. Is it not reasonable to suppose that muscles accustomed to daily toil for sustenance, when deprived of that healthful exercise, should become weak, flabby, and deprived of much of that vitality which constitutes perfect health? Departures from the irrevocable laws of animal life in its perfection is invariably accompanied with loss of some kind, and hence violation of physiological laws are dangerous.

We need not wonder that an active nervous system, from close confinement and relief from all anxiety about satisfying hunger, should change the temperament to a lymphatic one, which is the prevailing one of fat animals as a rule. We need not wonder that changes so conspicuous should lead to disease and a shortened span of life; that stamina of constitution and longevity should be wiped out with the sponge of disease. We conclude, then, that the above causes indicate a condition of the system which predisposes it to the taking on of certain diseases so fatal to the hog. These are, in medical language, the remote causes; the immediate causes now require a brief notice. The class of diseases which, under various forms, takes off so many hogs, horses, and cattle, has proved a stubborn enemy to veterinary students; and *post-mortem* examinations have only revealed the existing pathology of diseased parts, not the immediate existing cause of the phenomena presented. This class of diseases seem to belong especially to the mucous membranes, those tissues which are exposed to the direct action of causes existing in the atmosphere or in the food. The causes of epizootic diseases, and those which produce typhoid types of disease through the medium of the bowels and stomach, are floating in the air, or exist in the food taken into the stomach. It is now admitted by some of the best authorities that epizootic diseases are caused by a vegetable growth, the minute spores of which are breathed into the lungs, as they are floating in the air we breathe, and also that some typhoid forms of fever, as hog-cholera, are of either animal or vegetable growth, and that the spores or minute eggs are introduced in the food. What is singular to the non-physiologist, these spores coming in contact with healthy mucous surfaces will not vegetate, showing that certain, definite conditions are required in this membrane to produce disease at all; or, in other words, there must be a peculiar abnormal condition of this membrane before there can possibly be a development of these diseases. A further examination of the matter of the stomach and bowels by a powerful microscope is very desirable, that more positive and reliable knowledge may be gained, which may point out a treatment which, thus far, has been little less than an opprobrium to veterinary practice.

Symptoms of hog-cholera are not unfrequently modified, or new symptoms added. The characteristic symptoms, which are never absent, are fever, refusal to eat, disposition to lie undisturbed, and a fetid discharge of dark-colored faeces. We suspect the distinctive feature which shall distinguish hog-cholera from all other disease will be found in the peculiarity of the fecal discharges, and these can only be demonstrated by careful microscopic investigation.

The treatment upon which any reliance can be placed, so far as we know, has not yet been discovered. It is true quackery raises her hydra head, and floods the country with sure cures, but whether from medicine taken or in spite of it, we do not know. As a rule, about the time we find out the hog is really sick, the disease is so far advanced that remedies may be considered useless. We have seen it stated that turpentine has been given, about a teaspoonful to the hog, and with success. A further trial is desirable, for it is not impossible that turpentine may kill those minute specks of life without injury to the mucous membrane. An accidental discovery of this kind would save millions of dollars annually. But there are other diseases, among which pneumonia is not uncommon and often fatal. For instance, we have known cases where the hogs piled themselves up on the wet ground under cover, so that they became steaming wet; they then rush out into the cold air to eat their corn, take cold, and die of pneumonia. Hogs are often troubled with worms, which greatly disturb digestion and make the appetite capricious, keeping them thin in flesh. Copperas in their swill, at the rate of two table-spoonfuls to the pail of swill, will clean out the worms and greatly improve the health of the hogs. Repeat this twice a week for a few weeks. A large farmer in Kendall County this fall lost 300 head of hogs, but he came to the conclusion, whether the true one or not, that the disease was not true cholera, but a form of disease which he believes was produced by a stagnant pond of water in the field. They were in the pond a good deal, and the pond was covered with a green scum. This may have been a malarial disease in some respects analogous to the genuine cholera.

Since, from the nature of the case, the disease is not noticed until it is fastened upon the system and beyond the stage in which curative measures may prove successful, it is wisdom to fall back on a surer and more feasible plan—precautionary measures of prevention. The question arises, What may be considered in some sense prophylactics in this class of diseases? The answer is, Preserve a healthy play of the organs of the body, and the causes producing these diseases cannot act on the mucous membranes, and consequently no disease will be produced. A weakened and partially diseased mucous surface seems to be a prerequisite to the sprouting of spores in the lungs or the hatching of eggs in the stomach and bowels. Right here we are met with the very pertinent question, Can we prevent the development of disease where the predisposition is always present by any treatment of the animal? Like hereditary consumption in man, so long as the health of the animal is sufficient to resist the causes acting on the predisposition, so long will the disease be absent. What, then, can be done toward saving millions of hogs annually? First. They must have a dry and comfortable place to sleep, and this apartment should be cleaned out every few days, and, if necessary, washed out also. Second. They must have clean water so arranged that they can drink whenever it suits them. Third. They should have salt at least twice each week and stone or charcoal, which is better, every week. Fourth. They should be fed upon a clean floor, and their feed should be mixed or frequently changed; cooked food, with apples or potatoes for desert, and then corn in the ear or hasty pudding. Fifth. In summer they should have all the timothy and clover they will eat. This treatment would doubtless save a host; but so long as a predisposition exists, there will be more or less disease, and so long as new varieties are being developed, there will exist an instability in breeding, which tends to weaken rather than strengthen the constitution of the hog. We doubt seriously whether hog-cholera, under present modes of breeding, can be either prevented or successfully treated. Still, accident may discover a remedy which will kill the living cause of disease without injury to the animal. Of course we do not recommend going back to the "alligator pike" or the "Ohio rooter," charged with stealing potatoes out of the second row in the adjoining lot. We do believe, however, that the hog needs more exercise, a greater variety of food, and that he should not be bred in and in, as all our best breeds have been. We have too many varieties now, and the more we get and undertake to breed them pure, the weaker and more liable to disease will the hog become.

Mr. E. STOKES, Berlin, Camden County, New Jersey, says:

We have been exempt in a great measure from diseases among our farm-animals in this immediate vicinity for some months, except a disease affecting the horse. This malady is very fatal, and a number of horses have been lost in the southern portion of this county and many in Atlantic County. They are taken suddenly with great weakness, and in many cases very soon after eating a full feed are unable to stand, and in four or five hours become perfectly blind and experience great difficulty in breathing. They die within from twelve to twenty-four hours. Almost every case has proved fatal. Mares seem much more liable to be attacked than horses. I have heard of no mules being attacked by the disease. Horses in prime condition are as liable as those that are not, and young ones are rather more liable than old horses. I think the disease is somewhat on the decrease at this date. Some localities are entirely exempt, while it may prevail on almost every side. Should the disease become general, it will prove much more serious than any malady we have ever had among our horses.

Both hog and chicken cholera are prevailing to some extent in this locality.

Mr. J. C. THORNTON, Elliott, Ford County, Illinois, says:

A disease exists among hogs here which has proved very fatal. In the fall of 1875 I lost all but sixteen out of a herd of one hundred and twenty. The symptoms of the disease vary a great deal. The first symptoms are invariably manifested in a dry cough, great thirst, and sometimes purging and vomiting. As a general rule, hogs, while under the influence of the disease, are very stupid. The duration of the disease also varies. Some of those affected will linger along for a month or two; some will apparently get better, but after a while the flesh will begin to drop off in places, and then the animal will soon die. The larger portion of those attacked will die in a few days. I gave new milk from a fresh young cow to the first two of my hogs that were affected, and they got well; but I could find nothing that proved of any benefit to the others. I used stone-coal, copperas, sal-soda, sulphur, alum, cayenne pepper, &c., without any beneficial results.

The disease prevailed in an epidemic form, as hogs were attacked without coming in contact with infected stock. During the fall of 1875 at least 1,000 head of hogs died of the disease in this township, a tract of land only six miles wide and about nine miles in length.

In the fall of 1876 the disease prevailed again to a considerable extent, and many hogs were lost. The symptoms were about the same as those given above. During the past summer the disease again made its appearance, but this time in a milder form.

A disease called chicken-cholera has proved very fatal to fowls in this locality. The fowl, when attacked, becomes stupid, refuses to eat, and in a day or two will die. Sometimes the comb or gills will turn pale or white. As a preventive, we use copperas in the water or in the feed with good success.

Mr. THOMAS TASKER, Angola, Steuben County, Indiana, says:

This county has been comparatively free from diseases of farm-animals, with the exception of epizootic or distemper among horses. The disease made its appearance last July, and still prevails to a considerable extent. It is very difficult to contend with or manage. The horse is affected with a cough—something like distemper—but the irritation seems confined to the glands, and the disease appears similar to glanders. The horse will have the heaves to all appearances until the glands are relieved. It has proved fatal in some cases.

As a remedy, four ounces of chlorate of potash to one quart of water has been used with good results. A spoonful of this preparation should be injected into each nostril every morning and evening until a cure is effected. Some medicine that will act readily on the kidneys will also be found useful.

Mr. N. N. HALSTED, Newark, Hudson County, New Jersey, says:

In 1859-'60, the first year of the appearance of the pleuro-pneumonia in this State, I had the honor of being president of the State society, and, with Governor Olden's assistance and the generosity of some few of the members and officers of the association, we made an exhaustive examination into said disease; bought the diseased cattle, quarantined them, killed some and made, through our surgeons and veterinary surgeons from New York and this State, a careful autopsy of several we killed and many of those that died. The result of these investigations was published in the annual report of the State society. We went to Boston and made a thorough and careful examination there, and decided that the disease was an imported one.

The disease was brought into our State by Mr. Johnson, who bought six calves from the swill-milk stables in Brooklyn, N. Y. These brought the disease to his herd. The society stopped it there and we had no more of it until our Union County farmers bought some more swill-milk-stable animals, and, being sellers of milk, kept the matter quiet, or hid it from the officers of the society until the whole neighborhood was infected. This has been stamped out by a rigid quarantine and the use of carbolic acid, used as a disinfectant and by the animals inhaling it. They have some of it now in Burlington County, produced from the same cause, which is being eradicated by the same means.

Our society crushed out the Spanish fever by killing all cattle affected with it at the cost of the owners. All animals that die of this disease should be buried six feet under ground—hides, hoofs, and all—and the sheds whitewashed with quick lime and carbolic acid, as the disease is infectious.

Mr. Z. E. JAMESON, Irasburg, Orleans County, Vermont, says:

Hogs here are generally healthy, but during the past ten years there have been many cases of apparent paralysis of the hinder parts of young hogs, ranging in age from three months to one year old. At the present time a neighbor has three, about five months old, so affected. One of these cannot walk at all, one can only walk with his fore legs, and the other can use his hind legs but little.

These pigs have been kept in a pen 10 by 12 feet, with a plank floor, and fed almost entirely upon sour milk. Within a few days they have been allowed to run in a yard where they could have access to the soil, but no grass or green feed. No remedy is known. Some die. Others live until they are in tolerably fleshy condition, and are then killed for meat. The cause of this trouble may be in the lack of variety in food.

Mr. J. S. LATIMER, proprietor of Cedar farm herd of short-horns, Abingdon, Knox County, Illinois, says:

Diseases of horses in my locality consist in what are familiarly known by our quack horse-doctors as bots and epizootic or distemper, the first of which affects the horse internally. The remedies usually recommended and applied are too numerous to mention. Each doctor has a different one, and the remedies kill about as often as they cure the animal. No effectual remedy has yet been found, as a horse once affected with the disease never entirely recovers. The epizootic is a malady which affects the lungs and throat, and sometimes spreads to the limbs and body of the horse. We have what is known as regular distemper, which is of a milder form than the epizootic; but both are evidently the same disease. The quacks have different remedies, with none of which am I conversant. The disease attacks and destroys animals ranging from six months to two years of age. It seems to be contagious, and prevails at all seasons of the year. It is usually more fatal to older stock, as about 10 per cent. of those affected die, and those that do not are rendered comparatively worthless.

The cattle in this county, all along the line of the great thoroughfares, are subject to attacks of the Texas cattle fever. In this county we are annually subjected to it. Then we have the disease known as black leg, which is virtually a blood disease. It affects young stock principally, mostly calves from three months to one year old, and is very rapid in its course. The calf frequently dies within thirty-six hours after the first symptoms of the disease are observed. On skinning the animal, all the blood vessels of the legs and neck are usually found clotted and gorged with black blood. So far no remedies have been found. In certain localities in the county it is more virulent and fatal than in others. When a lot of stock is attacked it usually goes through the whole herd. It is very fatal, and I regard it as contagious. Perhaps 6 per cent. of the young stock of the neighborhood die of it. Usually the calves that are in best condition die first; thin ones are rarely attacked.

Another troublesome and growing disease is that of abortion in cows. The disease is little understood—indeed its causes are a mystery to us all. I believe it to be a blood disease, and under certain conditions contagious. When once started in a herd of cows, let them be ever so healthy, it is apt to affect them all. They lose their calves anywhere from three to seven months' time. Unless well cared for many of those affected will die, or if they do not they will afterwards prove worthless as breeders. I have tried, and seen tried by a great many others, various remedies, but all have proved worthless. Changing from one pasture to another, and separating the well from the affected ones, will sometimes do good for a short season ; but the disease will usually break out again, perhaps affecting cows that were previously exempt. The opinion generally prevails that the disease is contagious. For the past two years I doubt if 10 per cent. would cover the annual losses from this malady.

We annually lose at least 20 per cent. of all our hogs and pigs by a disease commonly called hog cholera. Many diseases are classed under this head, and some of them are no doubt the result of local causes, such as bad treatment, confinement in filthy and ill-ventilated buildings and pens, &c. Worms in the throat and intestines is one of the symptoms of the so-called cholera. Many specifics are used, but no certain remedy has as yet been found. Copperas, sulphur, charcoal, turpentine, asafetida, antimony, and many other drugs have been tried, but usually without satisfactory results. The disease is certainly contagious, and one of the best preventives is to separate at once the sick from the well hogs, and divide the well ones up into small herds. A change of feed from corn to oats, bran, &c., will also be found beneficial.

Mr. G. W. BALDOCK, Charlestown, Clarke County, Indiana, says :

The disease known here as hog cholera seems to prevail all over the hog-growing country. It prevails as an epidemic in this neighborhood and county. Mr. David Lutz recently lost one hundred and twenty-three head ; Mr. Isaac Koons two hundred head ; Mr. Floyd Ogden, two hundred head ; Mr. Samuel Lewman, forty head ; Mr. G. B. Lutz, 50 head ; Mr. John King, 50 head ; Mr. David King, 35 head ; the writer, 50 head ; and so on throughout the entire neighborhood. All diseases affecting swine are erroneously classed under one head—that of cholera. My hogs were afflicted with what I considered a lung disease, the symptoms of which were about as follows : The animal became very stupid, and lost its desire for food. It would mince slightly of its food, but would swallow but very little. Some of them would cough a great deal and others but little, while still a few others would not cough at all. Although the coughing showed the presence of disease, I did not consider it one of the leading symptoms. After the disease becomes fully developed they become constipated, and the faeces hard and very offensive. They nest around and seem to want to sleep all the time ; eat nothing and soon die. There is no known specific remedy for this disease, be it what it may. As a remedy I tried sulphur and copperas, wood ashes, and soft soap. These things seemed to give the well hogs a fine appetite. I gave one shoat a half pint of castor oil, which purged it freely and it recovered. As soon as I commenced feeding the above ingredients I had no more sick hogs. · Perhaps some of them may prove a preventive, but I am sure neither of them can be regarded as a remedy.

My wife has lost a great many fowls by cholera. We tried many supposed remedies, but without avail.

My neighbor, Mr. A. J. Crum, lost eleven head of cattle this summer by an unknown disease. They would froth at the mouth, quit eating, and soon die. He tried no remedy.

Mr. JAMES E. FOSTER, Brownstown, Fayette County, Illinois, says :

While we have lost heavily the past season with hog cholera, still I do not feel competent to give an intelligent diagnosis of the disease. I think there are two or three different diseases classed under the name of cholera. In the spring season the animals are affected with something like influenza. They cough and exude an offensive matter from the nose, refuse to eat, and pine away and die in from one to three weeks. Another and more fatal form is, I think, a typhoid or bilious fever. The symptoms are vomiting and sometimes purging. Those afflicted in this way die within a shorter time than the others, say within from one to six or eight days. The fatter the hog the more rapid and fatal is the disease. The percentage of recoveries in this form is very small. Remedies are attended with little success, as the animal is a hard subject to get medicine down. There seems to be no intelligent mode of treatment, and the trouble and expense often equals the value of the hog after recovery. I think measures of prevention will be found both more practical and more profitable. I would therefore suggest the isolation of the sick animals and the burying of the dead carcasses.

Mr. JOHN C. ANDRAS, Manchester, Scott County, Illinois, says:

In this vicinity the losses have been very great from diseases among hogs, that of cholera being the most prevalent. The loss of pigs recently, from one to two months old, within a circuit of two miles, has been over 400 head. In a herd of 150 head only two were left; in another of 90 head but 8 were left. The first symptoms were extreme chilliness, even when the thermometer ranged from 90° to 95° Fahrenheit. This was shown by their crowding in beds at mid-day, and a general discoloration of the skin, that of black hogs assuming a gray or purple hue, and the white animals a pinkish tinge. This was followed by high fever and a general breaking down of all the animal tissues, and fatal results within from three to five days. With older hogs the preliminary symptoms are the same, but the fatality is not so great. Recovery is generally followed by loss of hair and sometimes the sloughing off of large pieces of flesh. The animal is almost worthless for feeding purposes for at least one year.

As to remedies there have been none found that can be relied on with any certainty. Different compounds of antimony, arsenic, poke root, and iron (sulphate of iron), are used in some cases with apparent benefit. Dissection shows a general inflammatory condition, centering sometimes in the stomach, but more generally on the lungs. The general breaking up of all the animal tissues is shown by rapid decomposition as soon as death ensues. The usual bird scavengers seldom feast on the carcass of a hog that has died of cholera.

There are several other diseases which hogs are subject to, among which is pneumonia. The symptoms are high fever and general debility, and ultimately extreme emaciation, with small percentage of death. Long continued and the best of feeding will rarely overcome the extreme leanness of the animal. Dissection generally shows atrophy of part of the lungs, and general adhesions. I think a thorough investigation of this subject by competent persons would result in great good to the entire country.

Mr. P. T. GRAVES, Burkville, Lowndes County, Alabama, says:

All kinds of farm animals, with the exception of hogs, have been healthy during the past few years. Hogs have been affected more or less fatally each year for some years past with a disease known as cholera. The disease manifests various symptoms, the most fatal of which is purging. The excrement of the hogs affected in this way is of a greenish color and starchy consistency. No settled conclusion has been reached as to the cause of this malady, nor has a remedy been found. Two points, however, seem to have been conclusively determined, viz: First, that the disease commences in damp, warm, weather, during a favorable season for vegetable growth and fungoid formations. The hogs feed greedily on growing vegetation, with us mostly on cotton, and if allowed all they will eat the result is invariably disease. It is thought that atmospheric conditions have considerable influence in producing disease. Second, we find that hogs taken from a range where the disease has been developed, but showing no signs of infection themselves, if confined on dry ground and fed dry food they will escape the disease. But a clearly marked case of hog cholera is contagious, and the disease should be so treated. Those that have been so affected should never be used as breeders, as the taint will be imparted to the offspring. There are many remedies, so called, but caution and preventive measures will be found the most profitable.

All kinds of fowls have suffered to a great extent with cholera this year. Entire flocks of turkeys, geese, ducks, and the common barn-yard fowl have died from its effects. The disease is more fatal with the Asiatic breeds than with the more common kinds. No treatment has been tried with sufficient care to warrant a favorable opinion of its efficacy. Lovers of fowls and eggs will be grateful for a sure remedy for this scourge.

Mr. JOHN F. LAFFERTY, Martinsville, Clark County, Illinois, says:

I keep but few hogs, as the losses are so great that the business is not profitable. It frequently occurs that an entire herd is lost. While the disease is generally, almost invariably termed cholera, the symptoms are sometimes very different. For instance: Last summer my hogs first showed a lack of appetite, weakness in the back and a staggering gait, dullness of the eyes, general feverishness and great thirst. Finally they would fall down with a spasm, froth at the mouth, and squeal from the intense pain of cramping. The first stage would last from one to two weeks, but after the spasms set in, which daily increased in frequency, but three or four days would elapse before death would ensue. I lost eleven head out of a herd of nineteen with the above symptoms.

In August and September many farmers lost their fat hogs by what was supposed to be sore throat. They would refuse to eat, apparently because it hurt their jaws to masticate their food. In two or three days they would die, apparently without pain.

Chickens, too, are subject to a disease generally called cholera. I am of the opinion that the disease has its origin in the liver, as that part is usually found enlarged to three times its natural size. We often find the livers of apparently healthy fowls entirely too large.

I have tried all the popular nostrums and many of those little known for both hogs and chickens, but none do any good. We generally separate the sick hogs from the well ones and let them die. We kill and bury the chickens, or feed them to the hogs as soon as we discover any symptoms of the malady. I do not know whether the killing of affected hogs would arrest the disease, as I have not tried that.

Hogs and chickens are about the only classes of farm-stock affected in any way with disease. So fatal are the maladies which affect these that farmers have about abandoned both.

Mr. James T. Coleman, Collier County, Texas, says:

At this time we have no fatal diseases among farm-animals worthy of notice. At times we have had lung-fever and staggers among horses, and occasionally a few cases of malignant distemper. Cattle have suffered but little. Some cases of bloody murrain now and then occur, but so seldom that the subject is hardly worth noting. We have but few sheep in the county, and as far as I am advised no disease exists among them. Hog and chicken cholera prevails to a less extent than usual. All diseases affecting hogs and chickens, from time to time, are designated under the one head of cholera. The general symptoms in chicken cholera are about as follows: The comb and wattles turn pale, the fowl becomes droopy and stupid, the excrements are watery. Death ensues in a few days. Sometimes fowls that are in apparent good health will suddenly drop dead. On opening such the liver appears enlarged to three or four times its natural size, and is quite rotten. Copperas, calomel, red pepper, and tannin are used as remedies, and sometimes with good results.

Mr. Luke Teeple, Belvidere, Boone County, Illinois, says:

I have had no disease among my farm-stock except a disease known as cholera among chickens. It was very fatal, as my entire flock died with the exception of a few young chickens. I tried many remedies but all to no purpose.

A strange disease recently attacked one of my neighbors' pigs, shortly after they were weaned. They would be found sitting in the position of a dog. When disturbed, and often when they were not, they would start off on a run, and heedless of where they were going they would often dash themselves with great force against any obstruction that lay in their way. They would fall down, get up and stagger around awhile and fall down again, and then lay and pant as though they were tired and almost exhausted. At other times they would jump up into the air, and continue to do so until death would relieve them of their suffering. The pigs generally died within from four to twelve hours after the first symptoms were observed. As high as fourteen pigs died of the disease in one day. As a remedy, saleratus was used at the rate of one pound to twenty pigs. None died after the administration of the second dose.

Mr. J. C. Peak, Vera, Fayette County, Illinois, says:

There has been no disease in this section, of any consequence, among farm-animals, for some time past, except the so-called cholera among hogs. This disease appears at all seasons of the year, in hot and cold, dry and wet weather alike. It attacks all breeds, ages, sizes, and in all conditions, whether fat or lean. It appears in various forms, all of which generally prove fatal. Some seasons it is most prevalent among pigs and shoats. At other times these escape and the older hogs will be attacked, while during other seasons those of every age and condition will be suffering from it at the same time. The disease is generally preceded by a cough, sometimes low and suppressed and at others harsh and whooping. Sometimes the animal is costive and passes hard black lumps covered with white slime. Some will pass blood and also bleed at the nose. At other times the disease will assume the form of diarrhea, and the animal will purge severely and pass large quantities of black offensive matter. Internally we find the effects of the disease differing as widely as the symptoms. With some the lungs are found in a normal condition, while in others they are found diseased and decayed, as is also the liver. I have known instances where hogs would die very suddenly, and upon examination a shoulder, ham, or other portion of the body would be found bloodshotten and in some cases mortified. Last year I had forty head which seemed perfectly well one day, and on the next day they were sore, stiff, and lame. I lost thirty of them within as many days. I do not believe the disease contagious. I have known well hogs from other fields and farms to bed with diseased, dying, and dead hogs, and yet not become infected. Again, I have known those that were kept at a distance of a half-mile from diseased hogs and yet become affected with the malady.

Mr. J. BALLARD, Niles, Berrien County, Michigan, says:

Preventives for what is known here as hog cholera will be found better than cures. A great deal of this disease is produced by uncleanliness and a lack of pure water during dry seasons. Another cause is no doubt found in an exclusive corn feed. This food is dry and heating, and soon produces fever, which is one of the first symptoms of so-called cholera. If hogs are kept on good clover pasture, where they can have pure running water to drink and wallow in, with salt, ashes, and charcoal within their reach, and an occasional dose of sulphur, they will generally remain free from the disease. An occasional change of feed is always desirable, as but few animals will thrive continually on the same kind of food. The symptoms of the cholera are almost as various as the hogs themselves. Sometimes it will begin with a cough; one will appear lame in a hind-quarter, while another will bleed at one or both ears or at the nose; another will lose all its hair and bristles; another will eat heartily at night, in apparent good health, and will next morning be found dead. I have no remedy.

The symptoms of a disease affecting horses, known under the general name of epizootic, are a cough and loss of appetite, and soon a discharge from the nose. Rosin, saltpeter, ginger, and indigo are used as remedies with good results. The animal should be kept warm and comfortable, and given warm food of boiled potatoes and bran mash, or anything he will eat. Rub frequently and thoroughly, and give exercise, but not enough to heat the animal. The above remedies and treatment cured the worst cases we have had in this vicinity. Where strong medicines were given, several animals died, and others were a long time in recovering.

There is no prevailing disease among cattle at present. Occasionally we have a case of milk-fever among cows. A preventive for this trouble will be found in bleeding the animal a week or ten days before calving, and giving her a sufficient quantity of Epsom salts to thoroughly physic her.

Mr. W. P. COOPER, Alexandria, Calhoun County, Alabama, says:

The disease affecting horses in this locality for the most part is simply colic, caused by overwork and irregular feeding. All horses are more or less affected with bots, but they seldom attack until disturbed by an accumulation of gases. To prevent colic, moderate work, regular feed, and a proper amount of green food are necessary. If the physical condition of the horse is reduced disease will surely follow. As a remedy for colic, one ounce of chloroform to three ounces of sweet milk and one pint of whisky, mixed with one pint of water, and used as a drench at the month, will cure ninety-nine cases out of one hundred. As a remedy for bots, drench with one quart of lard oil. If not relieved in thirty minutes, repeat the dose. I have seen the bot die almost immediately when dropped in hog's lard. The grub breathes through the pores of the body, and when oiled they cease to breathe and death ensues. Nitric acid will not kill them, but oil will.

Native cattle here are subject to but few diseases, but imported cattle almost all die of a disease we call murrain. But few live to become acclimated. The symptoms are eyes feverish and excited; disposition to stand in water; very thirsty; discharges of bloody urine. In two or three hours the animal becomes uncontrollable and dies suddenly. On *post-mortem* examination one portion of the stomach is found perfectly dry. There is also found a large extended gall or bladder filled with bloody secretions. In the region of the heart are found collections of fluid which seems to be an overflow of bile from the gall. The disease is very fatal. We have no remedy.

Cholera is the only disease which seems fatal among hogs. When attacked the hog becomes stupid, its eyes matter, and it is often stiff and lame. Sometimes the animal is constipated and at others exactly the reverse. As a preventive, sulphur, copperas, salt, and strong wood-ashes in equal parts, mixed in slops, is given once a week. Cabbage leaves are regarded as an excellent food for sick hogs, and many believe them to be a cure for the so-called cholera.

Fowls are invariably healthy when kept clean. If the chick or older birds become lousy, tip the under feathers with grease and sulphur or mercurial ointment.

Mr. PERRY K. COLTON, Moorefield, Switzerland County, Indiana, says:

The only disease prevailing here among farm-animals is that among hogs, and known as cholera. There has never been a case of it in my neighborhood, but much of it has and does exist in adjacent communities. The first symptoms are langour, watering of the eyes, diarrhea, in some cases constipation, and a dry cough near the close of the scene. The duration of the disease is from one to two days. The average fatality is virtually all, for the few that do recover are afterward worthless. No remedies, so called, are used with any success whatever. Dissection after death discloses, in many cases, the bowels much inflamed. Often the intestines contain large numbers of white worms, which in some cases are so knotted together as to completely obstruct the bowels. The lungs are generally found much decayed and otherwise affected. Soap, black ammonia, wood-ashes, sulphur, &c., are given as preventives, but with what success would be difficult to determine.

We are of the opinion that the disease is a blood poison somewhat of the character of malaria. With us, where malarious diseases prevail in the human family, the cholera is mostly found, and where there is no ague or other malarial disorders there is but little or no cholera among hogs.

Mr. A. B. NICHOLSON, Lincoln, Logan County, Illinois, says:

In this (Logan) county, horses, cattle, and sheep are and have been very healthy. Hogs are afflicted by the so-called cholera. I am unable to give all the symptoms of the disease as they vary a great deal. Generally a loss of appetite, drooping ears, cough, diarrhea, &c., is observed. The younger hogs are generally the first attacked. There is not, to my knowledge, any known remedy. Very often a remedy is found and heralded over the country as an effectual cure, and it probably does cure some and then fails. The secretary of our State board of agriculture in March, 1876, sent out circulars containing forty questions relating to hog cholera, to upwards of one thousand swine-breeders. About two hundred and seventy were returned with the questions answered, but hardly two of them were agreed as to the cause or cure.

The treatment which is considered best is to change lots and sleeping places every week or two, with frequent changes in food. A preparation made of one bushel of wood-ashes, one quart of salt, one pint of sulphur, and one-half pint of black antimony should be mixed with their feed and given once a week. If your department can ascertain the cause and find a remedy for this disease, it will save millions of dollars annually to the farmers of the northwest.

Mr. J. A. JORDAN, Orion, Henry County, Illinois, says:

There is no special disease affecting farm-animals here except that affecting swine. What is known among us as cholera is at present and has for months past made fearful ravages among all classes of hogs. I am unable to furnish your department with the number of hogs that have died in my county (Rock Island) within the past four months, but after diligent inquiry I am satisfied that one thousand would be a low estimate of the loss we have sustained, and $15,000 would be a fair estimate of their value.

The cause of this disease is totally unknown, or merely conjectural. It is generally supposed, however, that it is caused by being fed too long in one place, or by eating their own filth. Feeding on plank floors and keeping them well cleaned off and sprinkled with slacked lime has proved highly beneficial.

Any description I might attempt to give of the hog cholera would be of little service to the department, as it is developed in a great many forms. I will, however, say that the hog when first attacked appears stupid and refuses to eat, is often very much relaxed, and occasionally passes what appears to be blood. They usually live from two hours to two or three days after the first symptoms are observed.

The breeding-stock growers here think that your department has never undertaken to investigate a subject so important to the people of the West, and indeed to the revenues of the government, as the one under consideration. I trust your efforts may be abundantly blessed in discovering the cause and a remedy for this terrible scourge.

Mr. GEORGE P. WEBER, proprietor of Meader farm, Pawnee, Sangamon County, Illinois, says:

The prevailing disease among farm animals and poultry in this section is known as cholera, and affects both hogs and poultry. Cases of Spanish or Texas fever among western cattle has in former years prevailed to an alarming extent, but for two years past I have known but little of this disease. The chief trouble being hog and poultry cholera, I will confine my remarks to these.

So much has been written and said on the subject of hog-cholera that its consideration has become almost disgusting. Nevertheless, in a work of such great importance, I am always ready to enlist. Swine, like all other classes of animals, are subject to numerous diseases; but since the first cases of what I regard strictly as hog-cholera were known in our county, all the swine ailments are called cholera. If an animal becomes affected in any way, the trouble being invisible, it is at once pronounced cholera. Hence, the great trouble so often encountered—incorrect treatment and ultimate failure. The disease was first introduced into this county about twenty years ago by large droves of half-starved Missouri hogs, bought there at a very low price, owing to scarcity of corn, and brought here to fatten when crops were fine. These animals were put upon a full feed of dry corn, and in a few days many of them were taken with violent fits of retching. In a few hours the bowels would begin to operate freely. Evident signs of griping in the bowels accompanied these discharges, which constantly grew more frequent and severe until death relieved the sufferer. Sometimes within a single hour from the first symptom the animal would die, while others would last twenty-four hours, or even longer. Very few of these animals, thus afflicted, recovered. No remedies that I have heard of were used, as it was thought to be caused by the high feed closely following the extreme starvation to which they had been subjected. In a short time, however, the native hogs began dying in a similar manner, which caused no little alarm. Since that time our county has not been free from this plague. Then began the discussions as to contagion, epidemic, &c., with which all are acquainted who have paid any attention to the disease. While these points have never been decided, I regard them as matters of great importance.

The symptoms of hog-cholera are about as follows: Disposition to remain quiet; when driven up to feed will smell of the food but refuse to eat; stand drawn up with feet under the body, back arched, head and ears drooping, eyes look weary and frequently inflamed; violent retching and vomiting; gripings and evident pains and cramps in bowels; severe scouring, and discharges not always of same character. Death usually ensues from within one to thirty-six hours. If the latter period is passed recovery is not unfrequent. Animals once affected are not so liable to attack in the future.

It would require hundreds of pages of closely-written matter to give in detail the varied treatment and remedies used for this malady. Almost all the minerals and vegetables in their different forms are prepared for medicines; stone and charcoal, lime and ashes, the different kinds of oils and salts, sulphur and soda and the various acids, mixed and compounded, mercury and arsenic; indeed the entire list is given for aught I know. We have known of seeming wonderful cures and strange failures under the same treatment and remedies. My opinion, founded upon practical tests and observations, is that the disease is epidemic and contagious. Animals should have the largest possible range; they should never be housed except in bad weather; their feeding-place should be changed as often as once in two or three weeks; their beds should be carefully attended to, and all the trash, old beds, and collections about pens and sheds should be burned as often as once a week, and the ashes left for the pigs to eat. Pigs should have access to pastures as much of the year as possible. They should be fed all the slops from the kitchen and the dairy, or as much of it as they will drink in the dry weather of late summer and in midwinter. Feed and water regularly, and never give medicine unless the bowels become constipated. Then air-slaked lime, wood ashes, and a little salt is the best remedy. The condition of the bowels may be readily known by watching the droppings. I am fully convinced that if the bowels are kept in a healthy condition there will be no such thing as hog-cholera, so-called, or in fact many other diseases. This should be done by cleanliness and careful feeding, watering, &c., and not by dosing with poisonous medicines.

Of course my *post-mortem* examinations have not been strictly scientific, as I am not a veterinary surgeon. The results invariably satisfied me, however, that the whole stomach and bowels were deranged, usually inflamed, as if greatly excited. I have found nothing that would justify a specified location, or a reasonable cause for the disease. I have examined many, as in former years I lost them by hundreds. After all my reading, observation, and actual experience, I pronounced the whole thing a mystery that can only be solved by accident, time, or science. Therefore I rejoice to see your department moving in the matter.

Chickens and turkeys of all ages are the principal sufferers from the malady known as chicken-cholera; yet other domestic fowls are not proof against it the disease. The symptoms, treatment, and results are so similar to the disease known as cholera among hogs that a full statement would amount to nothing more than a repetition of the above.

Mr. R. RICHESON, Ewing, Franklin County, Illinois, says:

While there have been some disases among cattle, horses, sheep, and poultry in the past that were the subject of some thought and investigation, their general condition and health at this time in this vicinity are such as to attract no special interest. Their health has been good, especially since the cessation of dry seasons and chinch-bugs.

With the hogs it is quite different. They are exceedingly healthy in all respects, with the exception of the prevalence among them of the disease known as cholera. From it no known condition, treatment, location, food, water, temperature, exercise, or season seems to give any guarantee of security. They take it at all ages and under all conditions, as people take measles or small-pox, and the surrounding conditions only seem to modify its effect in severity and fatality, the greatest effect generally being produced by the condition of the weather. In the mild weather of spring the percentage of fatality to those that take it is fully as low as 20 per cent.; in the fine weather of fall it is a little worse; but in the heat of summer it is often above 90 per cent., and quite as bad in the coldest of winter. Although it does not spread as rapidly during cold seasons, it makes very near a clean sweep of those that take it. The laws of its propagation are visibly these: The more the hogs are isolated the less liable are they to take the disease; the larger the herds, when it once gets among them, the greater is the percentage of cases, and in cold weather if one of those that bed with others takes it and it is not at once separated from those not affected the whole bed will take it and probably all die. The percentage of hogs that take the disease varies with the weather and other conditions, sometimes varying from 40 to 95 per cent. I have known a few instances of isolated herds, fenced away from any contact with other hogs, growing with perfect impunity through periods of its greatest ravages in the vicinity, which convinces me that the disease is a contagion, and is governed by the same laws of contagious diseases as those which afflict other animals. In this belief I have been strengthened by the fact that the great supply of hogs to the market come from those localities where there are no free commons for hogs and where the breeders raise and fatten their animals; also that the still-house pens, cattle-lots, and free common country, which used to raise the bulk of the hogs, are now the localities of the greatest devastation. If I am correct in the above views, the questions of diagnosis and treatment are merged into the one of isolation and prevention. I have often seen a complete diagnosis of the disease published, and any attempt on my part in this direction would necessarily be more tedious than profitable. I have noticed but few unvarying symptoms of the disease. These, somewhat modified in various cases, are: 1. A drooping of the head with a dull appearance. 2. A wheezing cough. 3. Falling away from the food. 4. A disposition to crawl under weeds, brush, or straw. 5. Redness about the ears and under side of the body. These are the only symptoms that are at all constant in the animal while alive; but some of them are now and then wanting, while there are a great many others of a varying and often conflicting character. After death, in the great number of cases that I have opened, there is one conspicuous feature, i. e., the absolute absence of blood in those that linger a few days, and the collapsed condition of the lungs. Otherwise, I never could find any evidence of either organic or functional cause of death.

The incipient stages and duration of the disease are as varied and irregular as other symptoms of the malady. I have seen hogs eat heartily at night in apparent good health and next morning be found dead. In most cases they will take a little food the first day, and sometimes for several days; again, they may live for weeks and finally die of the disease. The most general duration, however, seems to be from three to six days.

As remedies I have known almost everything being tried, both in the vegetable and mineral kingdoms. I have often heard of specifics, and known parties who believed in them, but it has invariably turned out that the cholera eventually got among their hogs under unfavorable circumstances of weather or other conditions, and they died as did those not treated with these specifics. I have doctored hundreds, and am satisfied that if I ever cured one that would not have got well without treatment it was with petroleum—drenching a hog of two hundred pounds with about one-half of a teacupful at a time once a day. But my experience is that if a hog has the cholera bad and recovers or is cured it has but very little value afterward. The only practical treatment is to change them to fresh quarters, separate the diseased hogs from the well ones, and isolate them from one another as much as possible.

In conclusion, I must express the opinion, which has grown to a conviction with me, that the only practical remedy for cholera is to isolate the herd, to prevent the moving of diseased animals through the country, and to prohibit their wandering about with impunity, carrying and spreading disease as they go. Leaving the matter to regulate itself has caused this locality, which formerly sent great numbers of hogs to market, to be short of a supply of swine to make meat for home use.

Mr. R. T. SMITH, Phillipsville, Erie County, Pennsylvania, says:

I am glad that an effort is being made by the government to discover the causes of the various diseases which from time to time afflict farm animals. The epizootic, which occurred some five years ago, has since annually affected horses in this locality, but not so severely as during that season of epidemic. My horses are more or less affected two or three times a year. They are just recovering from a very severe attack. They generally pass through all the usual symptoms of the disease, such as cough, swollen glands, running at the nose, sore throat, &c. One of them was so stiff that I could scarcely get him out of the stable. By allowing them to rest two or three weeks they generally come out all right, and get along well enough until the next attack. If you can discover the cause and a remedy for this and numerous other diseases existing among horses and cattle in this locality you will greatly benefit every one engaged in the breeding and rearing of stock.

Mr. W. W. HINMAN, Cambridge, Henry County, Illinois, says:

Hogs have been dying at a fearful rate in this part of the country for over a year past. The disease seems quite general and widespread. However, there are a great many farmers that as yet have had no sickness among their hogs. During February and March last I lost twenty-five head. The disease seems to attack the lungs, as a harsh, rattling cough is generally the first symptom observed. This is sometimes accompanied by vomiting and purging, the latter symptom being a very dangerous one. In most of the cases that came under my observation the animals were constipated. In all cases the excrement was very dark in color. There is nothing certain about the duration of the attack. Some die in a few days, while others linger for two weeks or more. I lost about one-third of my entire stock of shoats. Hogs that are nearly matured are not so apt to take the disease.

I do not know that I can give a diagnosis of the disease, as I have never been present when a *post-mortem* examination has been made. Of one thing, however, I am quite sure—the lungs are the place where the disease originates, and they continue to be the main cause of disturbance until the hog dies. I used various remedies, my first being wood-ashes and salt—two or three parts ashes and one part salt. After that I used turpentine given on coal (anthracite). This seems to help them. I also used carbolic acid, sprinkling the places where they slept and putting a small quantity into the water they drank. After using the carbolic acid thoroughly for a short time (two or three days) my hogs began to improve rapidly; in fact I think I lost but two or three afterwards, and they were bad cases when I commenced using it.

I hear of no complaints in regard to other kinds of stock. I do not know as this will be of any benefit to you, but "straws show which way the wind blows."

Mr. THOMAS D. OGDEN, Hoosier, Clay County, Illinois, says:

Horses, cattle, sheep, and mules are very healthy in this locality at present. Hog-cholera prevails to some extent. No remedies have been found that can be relied upon. If a sure preventive or a certain cure for this terrible disease could be found, it would prove a great blessing to the farming community.

Mr. WILLIAM B. STANTON, Pollard, Escambia County, Alabama, says:

Hogs here are afflicted with a disease called cholera. In 1874 I lost by this disease all the hogs I had but two, and they were worthless afterward. I kept one sow until she had pigs twice, and they all died within one or two days after they were dropped. The disease was very fatal, and often I would not know that anything was ailing the animal until it was found dead. Some lost but very few, while others lost nearly all they had. The disease has not been so fatal since. A few hogs have died every year, but the malady has not been so wide-spread as it was in that year. No remedy is known here. I do not know whether cholera is the proper name for the disease or not; I only know that that is what it is called here.

Mr. T. B. CALDWELL, Forrest City, Saint Francis County, Arkansas, says:

In our portion of country stock of all kinds receiving proper treatment are remarkably healthy. For the last seven years we have kept on an average seven head of horses, and during that time have not lost one from disease. We have had some cases of colic, caused by irregularity in feeding, and the epizootic in a mild form, neither of which required treatment. There has been some loss from charbon or yellow water, supposed to originate from impure blood, which proper treatment would perhaps have obviated.

Cattle are very healthy, with the exception of slight losses from murrain, which it is believed could be prevented by regular salting. This is proven by the seldom occurrence of loss where cattle have plenty of salt.

The loss from diseases among hogs is very great in this section of the State, but improved breeds which have good attention are healthy and prolific. Two years ago I had a couple of hogs afflicted with what was called "blind staggers." They appeared to have spasms, could not see, kept constantly moving about, and would sometimes fall as if dead. In about one minute they would get up and move off again, apparently relieved. Others of my hogs showed symptoms of the same disease. I had been feeding for some time on hard corn, and I think this was the cause, for when I changed their feed to bran mash they all got well. The two sick ones I saturated from ears to tail with coal-oil and turpentine. The losses in this county were about 60 per cent. during the above season from staggers and cholera. The cause of the last-named disease is unknown, but it is believed that proper feed would prevent it.

Fowls are subject to several diseases, the most fatal of which is known as cholera. The causes are unknown, as on some farms almost every fowl dies, while on adjacent farms none die at all. It is my belief that if properly cared for all farm-animals (except in cases of epidemic diseases) would be as healthy as could reasonably be desired

Mr. J. McGowan, Orland, Steuben County, Indiana, says:

On my own farm I have had no trouble with my stock, but my neighbors are suffering from the disease commonly known as "hog-cholera." From cases that have come under my own observation I am led to believe that the disease is more like typhoid fever as it affects the human family. On my farm we have fed eighty head of hogs with good success, and I cannot do better than give you our manner of handling them. We give ample range, with pure, fresh water constantly before them. Salt twice a week, and keep wood-ashes and lime continually within their reach. Our hogs are of the Poland-China breed, and are very thrifty and seem perfectly healthy. Farmers in this vicinity are suffering terribly from this scourge, and trust that your investigations may be crowned with success.

Mr. John W. Ross, Fitt's Hill, Franklin County, Illinois, says:

Horses in this locality are occasionally affected with epizootic diseases. The malady comes on without any apparent warning. The symptoms are generally about as follows: Glands of the throat swollen and distended, and limbs and feet swollen; contagious eruptive fever, with inability to eat or drink; morbid secretion of saliva, and decided constipation. In fatal cases the disease runs its course in from ten to fifteen days. Hygienic measures are about the only remedies resorted to. Warm poultices may with benefit be applied to the throat, and the bowels regulated with salts or sulphur.

The disease most prevalent among cattle is murrain. It is characterized by small vesicles in the mouth, on lips, gums, and tongue, with drivelings of saliva, often causing inability to eat or drink. These symptoms are accompanied with fever, swelling of the udder, and lameness. In fatal cases the animal generally becomes unmanageable, disregards the commands of the groom, breaks away and runs over the neighborhood perfectly frantic and furious. The disease runs its course within from three to ten days. Where animals are affected with this disease the bowels should be regulated by mild laxatives, and they should have comfortable lodgings, with soft, digestible food. As an application for the mouth and larynx, a mild astringent solution of half an ounce of alum, oxide of zinc, or sugar of lead to a quart of tepid water will be found beneficial.

The most formidable, and by far the most destructive disease of all, is hog-cholera. It often devastates the whole country of large numbers of swine. It occurs at any season of the year, but is generally the most prevalent in spring. The symptoms are a stiffness of joints, no desire for food or drink, dry, hot, harsh skin, general disturbance of internal viscera, nausea, and vomiting. The animal lingers from five to ten days, and generally dies. Decomposition takes place almost immediately after death. Various remedies have been tried, but with no decided benefit.

A disease known as chicken-cholera is also very destructive to fowls. When it breaks out in a flock it usually destroys the most of them. No remedy has been found to successfully combat the disease.

Mr. E. Stevens, Howardsville, Jo Daviess County, Illinois, says:

There is no special disease prevalent in this locality among farm-animals. Among horses the most troublesome complaint is that known as distemper or "strangles." It is quite prevalent throughout the low country of the Mississippi Valley. It is invariably known here as "distemper," and is of variable duration, often lingering for months, but seldom proving fatal to the full-grown horse. It attacks horses of all ages and conditions, and is highly contagious in its character. The disease is marked by three distinct stages. The first is a dry, hacking cough, attended by running at the nose. The discharge at first is thin and watery, and always of a whitish color. This discharge soon becomes thick and purulent; and the second stage rapidly follows by swellings or tumors under the throat along the salivary glands. These swellings soon establish an abscess in the throat, which rapidly enlarges until it breaks. This constitutes the third stage. If it breaks outside—which it generally does—matter may run for days or weeks, and sometimes for months, but the danger is passed if proper protection is afforded. But if this abscess breaks inside, the horse generally dies from suffocation or strangulation. The only remedy used here is the application of hot poultices about the neck and throat (in the second stage) to induce suppuration as speedily as possible.

Hogs are generally healthy, but when any die from any cause it is invariably attributed to "hog-cholera," when most likely no such thing as cholera ever existed among hogs in this locality. However, hogs frequently die here with quinsy and other throat diseases. The most successful treatment I know of is to give frequent small doses of powdered bluestone in sweet milk. I have also been successful by placing, with a wooden paddle, half a drachm of finely pulverized bluestone on the roots of the hog's tongue.

Fowls often die with what is known as "chicken-cholera." I know of no sure remedy. Equal parts of powdered charcoal and red ocher mixed with the food is an almost sure preventive.

Mr. F. M. ROGERS, Nora, Jo Daviess County, Illinois, says:

I have been a resident of this county for thirty-six years, and during that time have not known of any disease affecting cattle, horses, or sheep of an epidemic character. In many localities swine have suffered from the effects of cholera, but none so affected have ever come under my observation. Poultry has also suffered with the so-called chicken-cholera, a disease which often decimated large flocks in a few weeks. Of the various remedies recommended, orally and written, we have tried but few, and none of these with very gratifying results.

Mr. D. C. TOWNSEND, Fort Hill, Lake County, Illinois, says:

Last winter the sheep in these parts died by the hundred. There was nothing we could do that would save them. They would get dumpish and die in a few days. I lost one-third of my flock (one hundred and fifty). Since then I have been feeding them turpentine in salt, and they seem to be doing well.

At this time we are losing a great many hogs. They do not live over three or four hours from the time they are attacked. They turn black and bloat up. We are feeding them sulphur and charcoal; some give copperas. They will eat the sulphur and charcoal before they will eat their corn. I cannot tell what the result will be.

Mr. R. H. SAUNDERS, Pecatonica, Winnebago County, Illinois, says:

Stock has generally been free from disease in this county. There have been reported cases of the so-called hog-cholera this fall, but none have come under my observation. I have nothing to impart with the exception of the fact that my flock of sheep have been infested with parasites for some years past, causing a poorer condition and greater loss than formerly when affected in this way. Tape-worms in the intestines have been numerous, and have proven very fatal to lambs and sometimes to older sheep. Thread-worms in the lungs have also been numerous, as has a stomach thread-worm which resembles the lung-worm, but is rather longer. I think the cause attributable to keeping too many sheep on the same pasture for several years in succession. I have tried many remedies with but little success, and now consider a proper preventive the only protection. First, I change pastures as often as once in two years; plow and cultivate old pastures; do not allow them to drink of stagnant water; give them access to salt mixed with a little sulphate of iron. I have followed this plan for the last year or more, and have greatly reduced the losses. In corresponding with Professor Law, of Cornell University, on this subject, he states that the embryo of the lung-worm is found in the common earth-worm. It would be interesting to ascertain if such is the fact. If so, the cause is apparent, for as land becomes richer from sheep-manure earth-worms become more numerous. Professor Law seems to be uncertain as to how the thread-worm finds its way to the lungs. I will state here that I have found the stomach thread-worm in immense numbers in lambs not more than four months old, but have found no lung-worms until the lambs were seven or eight months old, and have found them most numerous in yearlings. Is it possible that they make their way from the stomach to the lungs?

Mr. A. M. DURKER, Howardsville, Stephenson County, Illinois, says:

Perhaps this locality is one most favored for the successful raising of farm animals, as there are seldom any prevailing diseases among any class of stock. Horses sometimes have distemper or throat disease, but in a majority of cases they require little or no treatment, as a very small percentage die, perhaps not one in a hundred.

Hogs are extensively raised in this locality, and I verily believe that no particular disease has ever prevailed here but what it could be traced to improper treatment and care. If a man loses many hogs it is attributable to hog-cholera, simply because they do not know what else to call it. I would here state that if the disease ever has prevailed to any extent in this locality there is no known remedy. I have never seen a case of hog-cholera that I know of since I have been a resident of this county, yet I have heard of cases attributed to that disease, and some farmers have been known to lose a heavy percentage. There are cases of influenza or pneumonia, caused by improper treatment and care in the colder part of the season. This is frequently brought on by keeping too many hogs together, and allowing them warm straw bedding, causing them to steam and sweat freely. In leaving their beds they cool off suddenly and take cold, which often produces congestion of the lungs, for which there is no known remedy.

Fowls are subject to chicken-cholera, but all the numerous remedies that have been applied have proved unsuccessful.

Mr. GEORGE STOCKS, Dalton City, Moultrie County, Illinois, says:

In regard to diseases among horses I will say that I have had more than my share of losses, but as I employed a veterinary surgeon, who, I think, rendered good service, I will not attempt a description of the disease, as I hardly feel competent to do so.

Hog-cholera is the scourge of Central Illinois. I have had some experience with it; I think it was in 1867, when I lost from thirty to forty head, all I had but one. The majority of my neighbors lost in about the same proportion. The disease was admitted by all to be the true hog-cholera. The animal would first commence to cough, would get off its feed, and its feet would seem to become very tender. It would creep to its bed with nose and tail down, and generally die within from one to three days. One widow woman near by lost none, and on inquiry I found that she kept a few ounces of asafœtida inclosed in a sack and suspended in the slop-barrel. I adopted the same preventive, and occasionally gave coal ashes, copperas, and sulphur, and for three years lost none. Early in 1871 I met with a report of a stock-grower's convention held, I think, in Lexington, Kentucky, at which one of the delegates stated that he put on the market every year from five hundred to six hundred head of hogs, averaging four hundred pounds in weight, and claimed that he could either prevent or cure the hog-cholera with the following prescription, viz: Four ounces of crystallized carbolic acid dissolved in one-half pint of rain-water. Dose, twenty-five drops to each hog, or one teaspoonful to four hogs, given in a little slop or milk.

On my place I generally have from forty to seventy hogs, large and small, and have used the above remedy for seven years with success. During that time I have lost only one, I think, and it did not have the cholera. Although this county has lost heavily we are not alone, as I yesterday heard of one gentleman residing near Dalton City who had lost twenty-three of a herd of twenty-six large hogs within the past few days.

I have given the above receipt to many persons, but often on inquiry have found that they failed to use it. Since using it I have had six or seven hogs so bad that they would neither eat nor drink, and I had to pour the medicine down their throats. In every case they recovered.

Mr. JAMES LILLY, Monticello, White County, Indiana, says:

Last fall my hogs were afflicted with a dreadful cough. Sometimes it was spasmodic with very difficult breathing. Matter was freely discharged from the nose and mouth, which was seemingly brought up by coughing. I usually fed them about the fourth or fifth day soft soap, and placed strong wood ashes in the trough from which they drank swill. With this treatment they recovered in from two to four weeks.

My neighbors' hogs had the cholera this fall, and they all died, that is, all that were afflicted. There is no remedy for this disease, so far as I know, that can be relied on with any certainty. It is believed that wood-ashes and soft-soap—in other words alkalies—are good as preventives. Probably this is owing to their tendency to cleanse the intestines of parasites. This would seem to indicate that the disease was caused by parasites.

Chicken-cholera prevails extensively and fatally at times, but I know of no preventive or remedy.

Mr. P. D. BOWLES, Evergreen, Conecuh County, Alabama, says:

The disease known as hog-cholera is characterized first by the animal refusing to eat, accompanied with slight dullness and sleepiness, which continues to increase from day to day, the hog all the time refusing to eat and hiding under the straw in his bed, where he remains for hours unless driven out. The feet refuse to perform their ordinary function of locomotion, and the animal limps or hobbles about as if there was a nail in each foot, back bowed, skin red, and after three or four days looks as if blistered; in fact the hair and skin finally all peels off of those that recover, leaving the animal almost nude. They eat very little for some days, but drink water in great quantities, and have copious discharges of urine, sometimes as much as a half gallon at a time, bowels costive. I do not recollect of seeing or hearing of a case of diarrhea or laxity of the bowels. The hog continues to decline, and either dies within from five to seven days or begins to eat and gets better.

The disease has prevailed in every township in this county to a greater or less extent during the past twelve months. It commences in the early spring and continues until late in the fall. It is generally more fatal among small pigs than among older hogs. I know one farmer who has prevented the malady from getting into his herd by giving "stack powders" two or three times a week in slops or meal. Although living in sight of his neighbor whose hogs died of the disease, his escaped. I was talking with another (No. 2) a few days ago, who said he had several pigs which he had kept penned up and fed on corn and slop, and that every one had died. Some man near by had a large number running in the woods, which were frequently turned in with those confined, but not one of them took the cholera. No. 3 had several pigs, all of which showed symptoms of cholera. He gave them a teaspoonful of spirits of turpentine in bran slop, and every one recovered. No. 4 has allowed all his hogs to run at large in the swamps, feeding a little corn at times to keep them gentle. Not one has been diseased. Upon general inquiry over the country I am prepared to say that all hogs that are allowed to bed in the woods and have free and large walks will escape the disease. Let him " root or die " and you will have no more hog-cholera.

Mr. John Powers, Rutledge, Crenshaw County, Alabama, says:

Hog-cholera, as it is generally known in this vicinity, prevails more or less every year. When attacked the patient begins to droop, holds down its head, and is indifferent to eating or drinking. They seem to be affected with a kind of dysentery, with frequent small evacuations. The surface is warm, and there are occasional quiverings of the flesh. Occasionally they die almost instantly. *Post-mortem* examinations clearly show that indigestion prevails.

The fatality is about 50 per cent. of those attacked. Generally three-fourths of a herd will be taken, while the remainder will continue perfectly healthy. Of those that overcome the disease about 50 per cent. regain their original health; the remainder are hard to fatten. I have never known a hog to die from the second attack. The disease prevails at any season of the year. Its fatality is greater among fat hogs, especially among those fed on corn.

The treatment is varied, but it is generally conceded that a small amount of alkali is the most efficacious, both as a remedy and as a preventive. A small amount of potash or concentrated lye is used by those who profess to treat it with any degree of success. We sometimes use asafetida as a preventive with success, but it is perfectly useless as a remedy. Corn feeding will not do; it will kill in nine cases out of ten. The hogs should be penned with shelter, free from dust, and sparingly fed on any easily digestible food. Whenever it is discovered that the animal has a desire to eat, be certain not to give it enough to satisfy it. Let it be kept hungry, not starved, but allowed about one-fourth the usual feed.

Fresh pine tar is good, both as a preventive and as a remedy, but it should be given in small quantities. Sulphur does harm, and copperas will ruin the teeth in a few days. Soda acts well. They require no external applications unless lousy. A lousy hog with the cholera would die if not cleansed. With the first treatment use alkalies perseveringly but sparingly, and the result will be 25 per cent. saved. Do not give corn unless it is ground.

Bots and glanders in horses, with an occasional case of distemper, are about the only diseases that affect this class of animals in this locality. Cattle are affected to some extent with hollow-horn, for which boring is the only remedy known here. Fat cows are never affected with this disease. A few cases of a new disease are reported, but enough is not known of it to attempt a diagnosis.

Mr. W. O. Millard, Caleta, Whiteside County, Illinois, says:

About one year ago the Secretary of the State Board of Agriculture of this State sent me a blank to fill out in regard to the so-called hog-cholera, which was then, as now, very destructive to all classes of swine. I made out a report and it was published in connection with a number of others from different parts of the State. When I made that report I had never been visited with the disease, and consequently was unable to give as accurate a diagnosis as I may be able to give you. The disease first made its appearance in this locality in August, 1876. But little attention was paid to it at first, perhaps because we thought it would not spread. But we were soon convinced that nothing had ever passed through the country that was so serious as this. It made its appearance in my herd about the 1st of June last, eventually almost annihilating them. When it first appeared I had two hundred and seven very fine animals of the best English Berkshire breed. Thirty days after I had but seventeen left, my loss being one hundred and ninety. While perhaps I may be considered one of our largest swine-growers, yet my loss was no greater in proportion than it was in the smaller herds.

The farmers all over this Western country are to day being visited with the worst scourge that has ever made its appearance. In this section they are losing from twenty-five to one hundred and fifty head of swine each. As to the nature of the disease I think it a typhoid fever, and it is so called by almost every one who has made an investigation. The first we discover wrong with the hog is its refusal to eat, and it acts, as we term it, dumpish. It either has a diarrhea or is costive. Its excrements are very offensive. Very many are taken with vomiting, while some are affected with bleeding at the nose. They seem to be thirsty and have a desire to lie in water a large portion of the time. Their eyes are red, and white matter stands in the corners of them, while many of them have a white mattery discharge from the nose. They usually live some two or three weeks after the first symptoms are observable. I have seen many of them where the fever had either settled in the head, eyes, nose, or legs, and in such cases some would become blind and others deaf. We have every reason to regard the disease as contagious, and I believe a prevention better than a cure. A few hogs recover from the disease, but a large majority die. We have done everything we could to effect a cure, but so far everything we have tried has proved a failure. I hardly think it necessary to say what we have given, yet it will do no harm. We have given arsenic, nux-vomica, calomel, salts, soda, concentrated lye, and Dr. Herrick's German hog cure. Bleeding has also been tried.

When my hogs were taken they were on grass, on a lot of seventy acres, well watered with pure spring-water, and had no grain. Others that were sick had grain and grass with good spring-water. Still others had grain and slops from the house and no grass or water; but all were sick. My land is rolling prairie, with no standing water or low places on the farm. The farmers generally are well off and take good care of their stock, and the majority have them sheltered in bad weather. The stock-growers here are very anxious that Congress should make an appropriation sufficient to investigate this matter thoroughly.

Mr. T. W. QUINN, Prattville, Grant County, Arkansas, says:

The only prevailing disease among farm-animals in this locality is cholera among hogs and fowls. Almost all the hogs in the neighborhood have been destroyed. Chicken cholera also prevails to an alarming extent.

Mr. L. H. COMPTON, Bay City, Pope County, Illinois, says:

The only diseases prevailing among any class of farm-animals here are those affecting hogs and chickens. So far as remedies go, there seems to be but little if any success in curing either hogs or fowls after the disease once takes hold of them. Every disease affecting hogs is called cholera, but my opinion is that there are as many diseases among hogs as "human flesh is heir to." Sometimes the symptoms indicate cholera, sometimes lung-fever, sometimes various other diseases, such as measles, quinsy, affections of the kidneys, liver, &c. I am of the opinion that many of these diseases are produced by worms, and in proof of the fact would state that those hogs that run at large and feed mostly on mast are the oftenest diseased, and these diseases are almost invariably caused by worms.

Mr. WILLIAM F. WATKINS, La Crosse, Izard County, Arkansas, says:

There never has been any scientific investigation in this country into the diseases of animals or fowls, and all the remedies used have been entirely empirical. I am not aware of any epidemic ever prevailing here among either horses or sheep. Our greatest losses are in hogs, which for many years have been (in different localities and at different times) subject to great fatality. The disease or diseases are confined to no particular season of the year, but rage only in certain localities at the same time. One locality of even a few miles in extent may suffer one season and be entirely exempt the next, while a neighboring locality is suffering. In a mountain district in the adjoining county of Stone, last summer, nearly all the hogs died. One farmer, by way of experiment, gave his hogs strychnia, both as a remedy and as a preventive, and lost but a small per cent. of them. Chickens and turkeys are attacked locally, just as hogs, but not in the same districts at the same time that hogs are. I have known calomel given to fowls as a remedy with very satisfactory results.

There has been a disease called the black-tongue among cattle in this county, which has only appeared at intervals of several years, and always in the heat of summer. It extends over the whole country, attacks only a small per cent. of cattle, but kills 50 per cent. of those attacked. It is more general and fatal to the wild deer than to cattle. It has not appeared for some years past.

Mr. L. Orto, Bradford, White County, Arkansas, says:

Hog cholera has been very destructive in some localities, yet I have been almost entirely exempt from the pest. I have kept from 100 to 800 head of hogs during the last ten years, and have had cholera among them but once. I then lost 80 per cent. of those attacked. When a hog is attacked by this disease the best remedy is to kill it and bury or burn the carcass, as this will have some tendency toward checking the spread of the disease. Moreover, if the hog should recover it will never be any account afterward. Hogs should never be allowed to sleep too long in the same beds. They should be changed about every ten days, and should be kept from dusty, dry places during the summer season. The oftener a hog shifts his range and bed the healthier will he be. They should have plenty of soap, lime, ashes, charcoal, and copperas. My hogs, which live entirely in the woods, are seldom affected with diseases of any kind. There are many wild hogs here, and I do not believe they are ever affected in any way. This is proof that the less this animal is hampered by close confinement the less is he liable to disease. The Poland-China and the Berkshire are the best breeds here.

Mr. J. S. Tait, Decatur, Macon County, Illinois, says:

I never lost any hogs until last winter, and I think that was the result of trimming in November and the early part of December. I then changed them from a warm bed to my cattle lot. Although this was covered and protected from the storms the ground was wet and frozen, and the hogs took cold and continued to drop off one by one until spring; but as soon as the sun came out and warmed up the earth they commenced to recover.

The only preventives I use are charcoal, wood-ashes, salt, and unslacked lime. In the summer season I put sulphur, copperas, and assafotida in the swill-barrel. I tie these drugs in a cloth and suspend it in the barrel. I give my hogs a roof to protect them from the storms of winter. If they have bedding it should be just sufficient to keep those on the outside from becoming chilled. Corn-stalks are the best bedding for swine.

My opinion is that hogs, as a general thing, are not properly cared for. Very often they become chilled through the night, or, if their beds are too warm, they take cold on leaving them early in the morning. Then follow lung affections, typhoid fever, and many other diseases to which they are subject.

Mr. William Dalgleish, Pleasant, Switzerland County, Indiana, says:

The disease known as hog-cholera has prevailed to an alarming extent for the past two years. I regard the disease as contagious. The symptoms are watering of the eyes when first attacked, followed by a dry cough, languor, thumps, constipation, &c. Death usually ensues within one or two days. The disease made its appearance among my own hogs in May last, and out of a herd of eighty-two I lost sixty-seven. I tried all the known remedies without any favorable results. From close examination and observation I am of the opinion that the disease has its foundation in the blood. The liver is generally torpid and the lungs much decayed.

Mr. R. M. Welman, Jasper, Du Bois County, Indiana, says:

There is no disease prevailing among any class of farm-animals except cholera among hogs.

Mr. M. M. Sloss, Simpson County, Kentucky, says:

We have never had any serious diseases among any class of farm-animals except among swine. Each of two distinct forms of disease destroys our hogs every year. One is called cholera and the other measles. The latter shows itself on the skin in sores and scabs. It is claimed by some of our farmers that sulphur, given internally, will effect a cure if it be given in sufficient quantities. Cholera is much the more fatal disease of the two. Generally the first evidence of its existence is the refusal of the hog to eat. Its ears will flap down over its eyes, giving it a dull, sleepy appearance. As the disease advances its breathing becomes hard and is accompanied with a symptom similar to thumps in horses. Usually its bowels are constipated, but sometimes they may be lax, and occasionally vomiting may occur. In a *post-mortem* examination of one case the entire intestine, and all it contained, were found to be very dry.

Few people attempt to cure a hog after the disease has taken hold of it. A number of practical farmers in this end of the county have used crude petroleum as a preventive for the past six years, and are established in the belief that if regularly and properly used it will keep them healthy. We buy it by the barrel, confine the hogs in a pen, and with a common tin sprinkler saturate them thoroughly from head to foot. We give it internally also on corn. Those of us who have tested its merits have great confidence in it, and in consequence have but little dread of the cholera. Where it has been used for six or seven years past the disease has not prevailed, notwithstanding its prevalence and destruction all around us. As I feel interested in the welfare of those engaged in agriculture, I hope you will pardon me for pressing upon your attention the value of the above article as a preventive of diseases among swine. I hope you will have its merits thoroughly tested.

Mr. HORACE J. LOOMIS, Chesterfield, Macoupin County, Illinois, says:

Many native cattle die here annually from what is known as Texas fever. The disease cannot be communicated except by Texas cattle, and they never have it. Cattle affected with it cannot give it to others, so there is no danger of its spreading and becoming an epidemic as many persons fear. The immediate cause of the disease is unknown here. The most probable theory with me is that it is transmitted by a poisonous substance in the urine of the Texas cattle. Whether any other cattle from Texas except the long-horned native breed can spread the disease I am unable to say, as no other breed has ever been brought here from that State. The subject should be investigated by scientific men.

Thousands of hogs die in this section annually. The disease assumes different forms in different localities, and in the same locality in different years. At times it appears to be a disease of the skin, and the hog will linger a long time before death ensues. Sometimes they will bleed to death from the nose in a few minutes, while to all appearances a few minutes previously they were well. In all its forms, however, there is more or less cough. I have examined many that have died, and in all cases have found large quantities of worms either in the throat or in the intestines. No locality appears to be exempt from the disease, and those who take most pains with their hogs are as likely to have the disease as those who are more careless. Hundreds of remedies have been tried, but as yet I have seen but little or no benefit from them. The whole thing is shrouded in mystery, and demands at the hands of the government an investigation by the most competent persons known.

Mr. CHARLES F. INGALS, Sublette, Lee County, Illinois, says:

Hogs are about the only animals subject to disease in our county, and so far as I have observed the ailment is of one and a similar type. It occurs at no regular intervals, and not oftener, I think, than once in ten years. I have been in the business here forty years, and until last summer my stock have kept comparatively healthy. Out of some two hundred shoats I lost about thirty, and those were the smallest and latest pigs. Grown stock seldom suffer. The animals lose appetite, become stupid, dwindle away slowly, and die, one here and one there as the case may be, in from one to three weeks after they are manifestly attacked. Upon being started up from their nests suddenly they usually are taken with a short hacking cough, but this does not continue when they are again at rest.

I do not now remember any stock-raiser who has twice had the disease to any extent among his hogs. Sometimes out of a herd of 200 head half of them will die inside of ninety days, and those that die first are generally the smallest. My usage is to give my animals extensive range, plenty of green feed, and to continually keep before them salt, ashes (wood or coal), stone-coal, and sulphur. They eagerly eat coal, and I provide it for them by the car-load. I have thought that high feed with Indian corn from generation to generation has worked constitutional debility in the hog. At any rate, after failing in finding any preventives, I have little faith in efforts to cure them after they once get sick. Isolation of all animals that are sick is found favorable to the well ones and to the recovery of those that are sick. Various specifics have been used and recommended, but so far as I know have effected but little good. If kept warm, dry, well fed, well ventilated, and in lots of fifty or less, disease will seldom be known.

Mr. WILLIAM BRINGHURST, Springfield, Utah Territory, says:

The climate and natural grasses of the Rocky Mountains are well adapted to stock-raising, containing elements that are health-producing and in their natural state an antidote for most diseases that stock are subject to. The epizootic, when raging here, was not fatal to animals running at large. The horse, however, when domesticated, is subject to two very serious distempers, which, if not promptly attended to, will prove fatal, the most common and serious of which is called the cramp-colic, produced by change of and over-amount of feed. The symptoms are restlessness, enlargement of abdomen, accompanied with severe pain. It will prove fatal in four or five hours. The most successful remedy used is one-half pound common sal-soda, two tablespoonsful of ground mustard, and one tablespoonful of cayenne pepper, mixed in water and given to the animal. The dose should be repeated in thirty minutes. Two doses are generally sufficient.

In Mountain Farcy, the cause of which is not known, the symptoms are a swelling under the belly, which extends rapidly over the whole body. I have seen the head swollen to such an extent that the animal was blind. It is very difficult to arrest unless taken at an early stage, and will prove fatal in a few hours. The remedy is bleeding in the neck. If the limbs are swollen bleed in each foot, striking the plate vein on the quarter between the hair and hoof. One-fourth pound of aloes, divided in three doses, as pills, or used as a drench, and given every hour, in addition to above treatment.

Horned stock has not been subject to any contagious disease in these parts, although there are isolated cases of hollow-horn, dry murrain, and fouls, which seldom or ever prove fatal. Cattle thrive well on the mountain grasses summer and winter, and require but little care. The raising of sheep is attracting much attention and has attained considerable importance, and under the management of scientific men is becoming very profitable. The Spanish merino is acknowledged to be the best adapted to this region. The only distemper in sheep that we are troubled with is the itch or scab. For this we employ the following remedy: After shearing dip the sheep at least every other year in a strong solution of tobacco and sulphur, composed of one part of sulphur to five parts of tobacco.

By experiment I find that swine can be raised profitably on the lowlands and on the borders of lakes and streamlets; but this class of stock are not generally bred here. They are not subject to any general distemper. The same can be said in regard to all kinds of fowls.

Mr. A. J. CARR, Charlestown, Clarke County, Indiana, says:

We are fortunately exempt at this time from any disease among farm-animals except cholera among hogs. I have lost a good many myself within the last two years. I tried all the remedies I could hear of, but none seemed to do any good. I then put some hogs that were apparently nearly dead in a close pen, gave them nothing to drink but a little sweet milk and soap-suds, with a little meal to eat, and they all recovered. My opinion is that none of the cholera remedies that are published as such and sold throughout the country are worth anything.

We have plenty of chicken-cholera, but so far no remedy has been discovered that seems to do any good. Preventives have been tried, but without beneficial results.

Mr. GEORGE W. THOGARD, Rutledge, Crenshaw County, Alabama, says:

I will endeavor to give you my experience with hog-cholera. In 1863 it made its appearance about thirty or forty miles south of this place. It then seemed to travel in a northern direction, and it took it near twelve months to travel a distance of 40 miles. Its destruction was at the rate of from 50 to 75 per cent. of the whole number of hogs attacked. There have been symptoms of the disease several times since without any marked direction as to its course of travel until 1876. In the spring of this year it made its appearance about 25 or 30 miles north of this place, and its course of travel was then from north to south. It took it six or eight months to travel south as far as this place. I can now hear of its progress south and west of here. On this visitation the average loss of hogs throughout the county was about 50 per cent. of the whole number.

I use lime, soap, salt, copperas, and blue vitriol as preventives, but my favorite prescription, and the one I believe to be the best, is poke-root and Jerusalem or wormseed root. I boil both together and mix the liquor with corn-meal while warm, and let the hogs drink it either cold or warm. The best remedy after the hog gets very sick is to kill it or have it removed from among the other hogs. The disease is more fatal and of shorter duration to fat hogs than to lean ones. Woods hogs are not so subject to the disease as those that run about the farm.

Mr. A. A. HOLCOMBE, V. S., New York, writes as follows concerning contagious pleura-pneumonia:

This disease was first seen in Central Europe about a century ago, and since that time has spread to most European countries, to Great Britain, Asia, Australia, and America. Its spread was undoubtedly due to contagion, for it is not at all probable that the disease originated spontaneously outside of Central Europe. It is a specific disease peculiar to bovine animals, for other species are never affected with it. It is always subacute or chronic in character; usually occurs as an epizooty or enzooty, and spreads easily and rapidly.

As the term indicates, the *lungs* and the *pleura* are the seat of the disease. It is not considered an inflammatory disease, and so far as local lesions are concerned, consists in an exudation of lymph into the connective tissue of the lungs, with effusion and exudation into the pleural cavities. The disease may be limited to one lung or it may affect both, while occasionally the pericardium is implicated. One attack usually confers immunity from subsequent ones. During its course the disease generates a specific virus capable of inoculating healthy animals of the same species with the same disease. By some few authorities it is believed that the disease can be generated by improper dietetic measures in conjunction with certain other influences, as excessive milking, and hot, illy-ventilated stables, but there is no positive proof to support this belief, although it is to be noted that the outbreaks in New Jersey in 1873-'74 and in 1877 were almost exclusively confined to cattle fed on beer-grains, which were kept in close stables, and gave large quantities of milk. The disease was brought to this country in 1849, and has prevailed to a greater or less extent in different localities ever since.

The period of incubation is reckoned at from twelve to sixty days, and the symptoms during this time are, as a rule, so slight as to receive little or no attention from owners or attendants. A rise of the bodily temperature is the first indication of the disease, and can be detected with the thermometer alone. Healthy animals have a temperature of 100° F., or a little less, so that a rise above this in an infected district would render all animals so affected liable to suspicion, for in those where the thermometer registers 102° F. or more the disease can almost positively be said to exist. The first symptom to gain the attention is mostly a short, dry, husky cough, of a peculiar character, and is first heard in the early morning, or while the animal is drinking. At the same time the appetite will be observed to fall off a little, and rumination be less active than common. The respirations are more rapid than normal, and may reach twenty, twenty-five, or thirty per minute, instead of about fourteen. Usually every respiration is accompanied with a low grunt or slight moan. The cough is growing more frequent, harsh, and painful; the back is slightly arched; the coat looks dead, and feels rough and harsh, while in some places it is erect; pressure along the back, especially in the neighborhood of the loins and in the spaces between the ribs, causes pain and flinching. As the appetite falls off the secretion of milk diminishes, until it is finally completely suppressed. The patient generally rapidly runs down in flesh, the surface temperature varies, the extremities being cold at one time and hot at another; sometimes but not always a slight discharge takes place from the nostrils, and the pulse becomes quite rapid. The lungs at this time are undergoing changes, easily detected by the expert; the air-cells admit but a limited quantity of air to the affected part; the intestinal tissue is filling up with lymph, and the pleura is undergoing the changes seen in this disease, presenting symptoms to be detected only by the practiced ear, as loss of the respiratory murmur, the presence of the different rales and the friction murmur of pleurisy, with finally the absence of any sound at all as the lungs become hepatized in the second stage, or the one of *marked* symptoms. In this stage the temperature increases and the pulse runs up to 60 or 70, and sometimes to 90, beats per minute. Examination of the heart will show it to be laboring hard to send the blood to the diseased lungs in sufficient quantity for the system; the extremities are cold; the front legs apart to facilitate respiration, which is becoming more and more rapid and difficult; the appetite is entirely lost; the secretion of milk has ceased; the feces are hard and dark colored; the urine is scanty and high colored; drinking causes hard and painful coughing. The animal almost refuses to move, seldom lies down, and stands with distended nostrils, moaning at every respiration, while from the eyes and nose is discharged a thickish, purulent fluid, and the breath is hot and fetid. These symptoms daily grow worse as the disease encroaches on the previously healthy lung-tissue; breathing is effected with the greatest difficulty; the pulse is so weak and small as hardly to be felt; the skin clings to the bones; dropsy beneath the chest takes place; the animal becomes almost unconscious of all surroundings, and groans and grinds the teeth; the abdomen fills with gas; diarrhea sets in, and death speedily closes the scene.

This is the usual course of a typical case where the disease runs through both stages and terminates fatally. In many instances there are variations from this general course, as where a fatal diarrhea sets in early or some other complication occurs which carries the patient off. (An interesting complication occurred in a case at North Branch, N. J., in 1874, where the lungs filled up rapidly and the pulmonary artery was ruptured.) But these variations are important only to the student of special pathology.

Regarding the course and termination of this disease, it is to be noted that it runs a more rapid course in young, vigorous animals than in any others; also that a short period of incubation is almost always followed by a rapid subsequent course. At times the disease terminates favorably in the early stage and before the extensive alterations of the lungs have taken place, yet these organs rarely regain their perfect function, part of their tissue ever after remaining impervious to air, while adhesions more or less extensive permanently exist between the lungs and the walls of the chest. The cough usually remains for a long period of time, being due to the alteration of lung-tissues. Death, as a rule, takes place in the second stage of the disease, and is due to the encroachment of the exudate upon the respiratory surface of the lungs, to anæmia, to gangrene of the lung-substance, or to a fatal diarrhœa.

The percentage of deaths which occur in the early part of an outbreak generally reaches from 60 to 90 per cent. of those infected, while later on, when the force of the infecting virus seems to have expended itself, the mortality may fall to 15 or 20 per cent. But this is not all the loss to which the infected district is subjected. The animals that recover are of little or no value for weeks and months, the secretion of milk does not return for a long time, and it is almost impossible to prepare them for market, for they do not thrive. Besides this, unless the subject of disinfection is understood, and its necessity thoroughly appreciated, all new animals are liable to take the disease and thus perpetuate indefinitely this dreadful scourge.

The intimate pathological anatomy of this disease, and the microscopical appearances of the involved tissue, can hardly be of value to the public, or to others than those thoroughly acquainted with histology, so that unless the department desires especially to have such, I will refrain from occupying your time with what can hardly prove of interest. I will therefore call your attention to the means of diagnosing this disease. The cough is peculiar, and to those acquainted with the disease would be almost sufficient evidence of the presence of the contagious form of pleura-pneumonia. The thermometer is of the utmost value in detecting the disease early. A physical examination of the chest, the temperature, character of pulse and cough, will always be sufficient to diagnosticate the presence of pleura-pneumonia. That it is contagious will be seen by the incubative stage, by the insidiousness of its course, and from the fact that it has no connection whatever with the causes which produce the ordinary form of this disease, that is, with climate, exposure, change of weather, food, &c. Also from the fact that it spreads by contact, and is very fatal. Lastly, some animals are not susceptible to the disease, about 15 per cent. escaping infection even when subjected to the influence of the contagion. The infecting principle of this disease is no doubt both fixed and volatile, for it is found in the blood, excretious, secretions, exudated lymph, and in the expired airs. The vitality of the virus is great, lasting sometimes for several months. It may be carried by the air a distance of at least three hundred feet, while by means of diseased meat, affected clothing, hay, straw, cars and steamboats, it may be carried to long distances.

Mr. R. B. DUNLAP, Boligee, Greene County, Alabama, says:

We are troubled in this section with two diseases among farm animals, both of which are very fatal. One is known as hog-cholera and the other as the "negro disease." It is hard to tell which is the most fatal to this animal. Remedies do not amount to much, and preventives will be found the most profitable and economical. I believe hog-cholera can be cured after the hog gets sick, but it is too tedious to have to drench them. I have cured a few cases by drenching them with the following prescription: One gill linseed-oil and one tablespoonful of spirits of turpentine. I generally keep a trough under shelter in which I keep about one bushel of hickory-wood ashes, one pound of sulphur, one-fourth pound of assafetida, one bushel of well-beaten charcoal, and a sufficient quantity of salt to make them relish it. This will not only keep off the lice, but will also keep the bowels in a healthy condition. Lice are the forerunners of cholera. They irritate the skin of the hog, weaken it, and render it liable to the attacks of this disease.

MORRIS CROHN, V. S., Erie, Erie County, Pennsylvania, says:

Since my residence here I have not observed any epidemic proper, though the splenic fever has been raging quite violently among the cows for the past month or so. Thus far it has been only local; and it is very extraordinary that, in view of the lamentable lack in this country of proper provision against the spreading of disease, the splenic fever has confined itself to one locality.

Splenic fever is due to the decomposition of blood; and, as the spleen contains a greater percentage of blood than any other organ of the body, it is most severely affected and is totally destroyed if the disease be not arrested. Besides this, the kidneys, and sometimes the bladder, will suffer from sympathetic affection, a bilious condition being indicated by the eye. I think that splenic fever has its origin in one locality, caused by dry pasturage, stagnant water, filthy stables, miasmatic air, and gaseous exhalations of the earth; and its spreading is due to the disease-matter in the air and immediate contact with infected animals. In every contagious disease there is a vital process, therefore all the properties of such a process are requisite for the existence of the disease. In order that the process of disease in an individual may develop, there is necessary the union of a predisposition (the inner element of disease) and an infection (the outer element). The predisposition, as the basis capable of development, is analogous to the conceiving function in the female, and the infection corresponds to the fecundating function in the male. As all disease is dependent upon the destruction of the healthy process, so the principle of disease in splenic fever is due to the unhealthy, abnormal condition of the blood, causing the decomposition of the latter and speedy death.

I think it incontrovertible that the decomposition of blood in splenetic fever may be accounted for by an insufficiency of iron in the blood. Proof of this is that in many cases coming under my personal observation, where a timely treatment with preparations of iron, together with tonics in emulsions, was pursued, the diseased animals were saved. One ounce of muriatic acid and fifty ounces of water administered once every hour, and after the fourth dose from one to three drams of quinine, is a very successful remedy. (Quinine, however, is too expensive for this purpose; cortex chinæ, from one to two ounces to the dose, may be substituted.) In addition to this, ice-water applications about the head and horns are of great benefit.

The disease appeared under three forms, with symptoms as follows:

1. Eye dull and inflamed; lack of appetite; feces thinner than usual, and slightly reddish; urine natural; pulse low; pulsation of heart increased. When the disease takes a fatal turn, chills and tremors appear; head and horns become hot; feces and urine bloody; pulse slow and at times suspended; beating of the heart perceptible; the eye assumes a dirty yellow appearance; horns grow cold and death takes place. Duration of sickness, six, forty-eight, and ninety-six hours.

2. Eye assumes a dirty red; pulse slow, suspending at times; beating of heart perceptible; urine bloody; feces similar to rice-water, offensive odor; head and horns hot. Duration of sickness from four to twelve hours. In this form of the disease a compound of one dram of opium and two drams of quinine has proven very beneficial.

3. I also noticed other varieties of splenic fever, which, however, were attended by no dangerous symptoms. Calamus and gentian combined with tannin makes a very good remedy.

There is a preventive to splenic fever used in Germany with good results, consisting of *natri sulphurici puto*, 540 grains, (*libram unam et dimidiam*), *sulphuris depurati puto*, 180 grains, (*unc. sex*). This is given in tablespoonfuls with the food.

There were but few non-contagious diseases which in their acute form have caused any serious loss. I will only mention colic and quinsy among horses and calving-fever (puerperal fever) among cows. Most owners of horses know nothing of medicine, chemistry, &c., but with the aid of "receipts," so-called "doctor books," and the advice of unqualified persons, they regard themselves as fully competent to "doctor" their horses. They almost invariably treat quinsy for glanders. They set up some arbitrary, wrong diagnosis, and give the poor animals large quantities of useless, injurious medicines, thus causing the loss annually of thousands of horses which were simply suffering from colic. These self-dependent men cannot tell whether the colic is caused by inflammation of the bladder, spasms of the bladder, suppression of the peristaltic action, gases, peritonitis, enteritis, &c., or by mechanical or organic obstruction; they invariably administer the same medicine rather than go to the expense of a rational veterinary treatment. Just so it is with puerperal fever, which, if not rationally treated, is almost always followed by death. During my practice here I have not lost a single horse afflicted with colic, or a cow having the puerperal fever, and therefore regard the remedies applied by me in these cases as *specifics*, which I shall only give to the public for a suitable remuneration.

Mr. THOMAS B. LUCAS, Easton, Mason County, Illinois, says:

Hog-cholera prevails to a considerable extent here. The disease makes its appearance about once a year. My hogs have often been afflicted with it, but never twice alike. The fatality ranges from 10 to 75 per cent. Remedies are numerous, but none seem to be of any account. A frequent change of diet and of range would seem to be the best preventive, and a separation of the older from the younger hogs. The disease appears to be more fatal along streams and in timber-lands than elsewhere.

Chicken-cholera also prevails to a considerable extent, with a fatality ranging from 10 to 100 per cent.

Mr. L. B. THORNTON, Tuscumbia, Colbert County, Alabama, says:

Horses here are subject to several different diseases, such as spavin, fistula, blind-staggers, glanders, &c. The best remedy for glanders is to shoot the animal as soon as taken, for the disease is incurable. Feeding horses on more oats and fodder and less corn will be found a preventive for many of the diseases which afflict them, and will also keep them in good condition.

Cattle are subject to murrain, and in most cases the disease proves fatal. Good pasturage and regular salting are also good preventives of diseases among cattle. Common colic in horses and cattle is generally cured by carbonate of lime, a teacupful in a pint of water and used as a drench.

For bots in horses I have found the best remedy to be a strong sage-tea, with molasses.

Hogs are afflicted here with a disease called cholera, for which no remedy has as yet been found. Calomel is extensively given. My experience leads me to believe that if hogs have plenty of good water, and are salted regularly and given sulphur and ashes and a supply of bituminous coal occasionally, they will escape many of the diseases to which they are subject. A great many of the diseases which afflict swine are caused by worms and lice. I use grease and tar for lice and calomel for worms, with good results. Last year, when the hogs were dying here by scores, I kept mine up in pens, with plank floors a little elevated. I kept the pens clean and used coal-oil and sulphur to destroy the lice. I kept a constant supply of wood-ashes and coal in the pens, and during the prevalence of the disease I did not lose a hog. They thrived and fattened well, and contained no intestinal worms when killed.

Fowls are subject to chicken-cholera, which is seldom cured. My experience is that if fowls are kept in clean, well-ventilated houses, and given sulphur and lime to keep off vermin, and are fed well, they will remain exempt from disease. Care and attention to feeding well are indispensable to healthfulness in fowls as well as all kinds of farm-animals.

Mr. O. E. LOVETT, Saint Elmo, Fayette County, Illinois, says:

We have no diseases among farm animals except among hogs, and with us all the diseases affecting the hog are classed under one head, that of cholera. Four-fifths of all the hogs in this county have died during the past summer. The disease presents itself in a variety of forms. Some become stupid and have a high fever. Occasionally they have a swelling on the jaw, shoulder, or hip; some on one part of the body and some on another. These swellings are generally filled with water. The bowels of many were not affected, and to all appearances were in a healthy condition. Hogs thus affected would live from four to twelve hours. Those that had very high fever usually become lame in one or more of their legs. After death their lungs were found filled with froth and blood, but the bowels were apparently unaffected. Animals thus afflicted generally lived from two to four days. Others that had high fever were bloated in the bowels, and would cough and purge when made to move about. They also passed bloody water. Affected in this way, they would usually live from two to ten days. About one-third of this class recovered and are doing well. I tried about all the remedies mentioned in the newspapers, and used lime and carbolic acid both in powders and in a fluid state as disinfectants, but I cannot say that either did any good. In July I lost about 100 head of hogs. They all died inside of three weeks. The disease appeared and disappeared very suddenly. The few hogs that recovered are doing finely, and appear as healthy as though they had never been sick.

Dr. A. JONES, Centreville, Montgomery County, Arkansas, says:

We have had no epidemic disease among fowls for some time. The past spring, however, we had some chicken-cholera, a disease characterized by a lax condition of the bowels, with greenish-white discharges, and a greatly enlarged liver and dropsical condition of the heart. I examined many and found all to be in the same condition, more or less, according to the advanced condition of the disease. No certain remedy was found, though many were tried. The best remedy, however, was ground mustard in salt dough.

Hog cholera, or a disease characterized by a high fever, nervous twitching of the muscles and slight cough, some looseness of the bowels, drooping, &c., has existed, more or less, for some years past among swine. Nearly all that are attacked die. Twelve months ago three of my hogs were attacked by the disease. A neighbor sent me word to give them one-half pound each of flour of sulphur, and he would pay for all that died. I did so, and they all recovered. Since that time I have had no more sick hogs. I would like to hear of the sulphur remedy being extensively tried.

Mr. W. N. COWAN, Gadsden, Etowah County, Alabama, says:

Cholera is the only prevailing disease among our hogs. Frequently its fatality is 50 per cent., and sometimes as high as 75 per cent. Various remedies are used, but with little or no effect. Epizootic in horses prevails to some extent, and in aggravated cases seems incurable. Some mild cases pass off like mild attacks of distemper. At some seasons and in certain localities cholera is very fatal among fowls. We have no remedy. Our ladies would rejoice at the discovery of either a preventive or specific for this scourge.

Mr. H. S. Dodd, Dodsville, Marion County, Arkansas, says:

During a residence of six years in this county I have not known anything like an epidemic among farm-animals or fowls in my neighborhood. In the county of Boone, adjoining this on the west, some cattle have recently died of what is called dry murrain, and many hogs have died of cholera. I examined one cow, and found the same symptoms present as observed in cases of Spanish or Texas fever in cases which I had examined seven years ago in the State of Kansas. I find on inquiry that Texas or southern cattle have been driven through Boone County the past summer, and therefore believe the disease to be the same. The first symptom noticeable is a sluggish movement. In the second the ears and head droop, the eyes sink in the head, and the toes of the hind feet drag on the ground. The duration of the disease is from two to six days. On examination the urinary organs present a very large and inflamed condition. The stomach is discolored to a black or dark red, and the contents are very dry and hard. Some remedies have been administered, such as diuretics and very active cathartics, with considerable success. My opinion is that such treatment is wise, and will in almost every case effect a cure where the treatment is persevered in and not delayed too long in the beginning. The diuretic used was nitrate of potash, and spirits of terebinthina the cathartic. Hog's lard was also used in large and frequent quantities. Congress would do a wise thing by making an appropriation for the investigation of these dreaded diseases.

Mr. H. F. Schenck, Cleaveland Mills, Cleveland County, North Carolina, says:

The only fatal disease we have to contend with, and which seems unmanageable, is what every one here calls cholera among hogs and chickens. It appears almost every summer or fall among the hogs in this county, and goes through one neighborhood one season and some other one the next. It does not seem to spread widely over the same country any one year, but seldom fails to appear each year. The animal when attacked first shows symptoms of drooping, and although they eat at first they often vomit after eating. They generally die within from seven to fourteen days. I would roughly estimate the fatality at 33 per cent. of those attacked. Of those that survive it is often two months before they finally recover. Many remedies have been tried, but with but little success. It has never but once attacked my stock, and therefore my experience with the disease is limited. Last year I had but seven hogs, and six out of the number were attacked. I observed it closely for a few days, and came to the conclusion that instead of its being cholera, as it was called by my neighbors, it was nothing more than simple constipation. They had no action of the bowels that I could discover. I gave them large doses of calomel, put them in a lot until it operated freely, and then turned them out. They all recovered. I advised my neighbors, whose hogs were similarly affected, to try calomel, which they did, and since that time there has scarcely been a death.

All that I have said about hogs is applicable to fowls.

Mr. J. F. Sellers, Perryville, Perry County, Arkansas, says:

Cattle a few years ago were subject to murrain, but now this disease is almost unknown. Horses have no diseases, except now and then a case of blind staggers, which farmers say generally arise from feeding inferior corn, and such diseases as fistula, &c., which are too common and the treatment too well known to require notice here.

We have at this time a disease raging among hogs which is thought by some to be the common hog cholera, but by others this is denied. The attack is made known by the general drooping appearance of the animal and a laxity of the bowels, though this last symptom is not seen in all. They generally die at the end of a few hours, greatly emaciated. They sometimes very suddenly swell under the throat after death, and these are thought to have been in some manner choked or suffocated. It has been noticed that those hogs that stay around houses and sleep in dry beds are much more liable to this disease than those that run in the woods and sleep without shelter.

Mr. James H. Rumbough, Warm Springs, Madison County, North Carolina, says:

Among some farmers of this section cholera sometimes prevails to the extent of destroying all the hogs on the farm. I have, however, never had a case, using as a preventive a weak solution of concentrated lye. I cannot learn of any intelligent remedy that is employed in this immediate section, and, having had no experience myself with sick hogs I am unable to suggest a remedy, or present any peculiarities of the disease, as I am not at all acquainted with the symptoms of hog cholera. But I am of the opinion that the disease in this climate is solely attributable to want of proper care and intelligent attention, over and irregular feeding, exposure to inclement weather, filthy quarters, want of salt, in the absence of which latter the animals sometimes resort to dirt and the accumulations in their pens.

The chicken cholera is sometimes prevalent here among that class of fowls which is the staple poultry of this section. I have no experience of any value in regard to this disease, and no suggestion beyond the want of proper care and attention on the part of a rustic population who have no idea as to the importance of attention, proper food, protection from the weather, provision of proper gravel, or cleanliness of roosts and quarters. Being a country of prolific vegetation, and the fowls being allowed to run at large over the farms, the young ones are subjected to the damp and cold of the dows and rains, which superinduce diseases peculiar to young chicks.

I am of the decided opinion that in a climate like this, naturally free from epidemic diseases to man and beast, that care and attention, intelligent regard to the comfort and food of animals, will constitute good, effective, and sure preventives of diseases of all kinds among animals and fowls.

Mr. W. H. Silow, Bay Minette, Baldwin County, Alabama, says:

Chickens have been affected more or less with a disease known in this locality as cholera. The fowls may be fat and perfectly healthy one day, and the next morning be found dead under their roosts. Some linger longer, droop around and gape a day or two, and then die. The gills become sallow and apparently bloodless. The discharges are green and very offensive. Not more than one in twenty-five recover. Some few have a second attack of the disease. Many remedies have been used, but I cannot say that any of them have proved sufficiently beneficial to be recommended. The disease is confined mostly to the Brahma breed. We have come to the conclusion that it is useless to doctor a chicken where the disease has progressed to any considerable extent. Black pepper appears to be about as good a remedy as anything else.

During the winter of 1876-'77 about half the sheep of this neighborhood broke out with what farmers called the rot. They lost flesh until they were greatly emaciated, and then the wool would almost all come off them. When attacked they would wander off singly or two or three together to some retired place, where they would linger for about a week and then die. I think about one-half of those attacked died. The loss caused a great falling off in the yield of wool in this section. They now seem to be doing well and are comparatively healthy.

Mr. Andrew Jay, jr., Jayville, Conecuh County, Alabama, says:

The importance of the object sought to be accomplished in your proposed investigation of the diseases of farm-animals would be hard to overestimate. It is a much-needed movement, for I know of no reliable remedy for any of the diseases which afflict farm-animals. The diseases existing among horses are colic, bots, or grubs, scours, staggers, distemper, and glanders. That among hogs is called cholera. Whatever disease may afflict a hog it is called cholera, yet it is very evident that the symptoms and effects greatly vary. Half of all the hogs in the county have died of some disease during the present year. Cattle are less subject to disease than any other class of animals. Occasionally, however, they have what some call murrain. Sheep likewise are subject to disease, and more so when huddled closely together. But I am too ignorant on the subject of diseases, as seems to be the case with all of our people, to have yet discovered or learned enough about the causes or cures to be of any real value. Cures are generally accidental, if at all. Sometimes the animals will recover in spite of the remedies given.

I regret that I am unable to contribute anything toward advancing so valuable and important a work. I would most cheerfully do so if I could, for I need its advantages and will be very grateful for any information growing out of this investigation.

Mr. C. H. Jernigan, Enon, Bullock County, Alabama, says:

Horses, cattle, and sheep here are subject to the usual diseases incident to these animals, for which various remedies are used. Hogs are subject to cholera, for which no remedy has been discovered. I would like to investigate this disease for the purpose of discovering its cause had I the means at hand. Chickens are also subject to cholera, so-called, and are also frequently afflicted with a disease called "sore head." No remedy is known for the first. As a local application, kerosene oil and lard, in equal parts, has been found a specific for the latter.

Mr. Thomas Dunnington, Pine Bluff, Jefferson County, Arkansas, says:

Notwithstanding the frequent prevalence of chicken cholera we find the raising of fowls profitable. The symptoms of the disease with us are a drooping appearance, indisposition or inability to eat, and death in a short time. We have used soda and sulphur as remedies, and cleanliness of houses as preventives, and by such means have managed to raise chickens and eggs sufficient for our own use, which we find cheaper than the raising of pork on the same amount of food.

In this section of country a great many hogs have been lost by a disease called cholera. It makes but little difference as regards symptoms; all hogs that die are afflicted with either cholera or mange. Those that are affected with mange are covered with a dry scuff, waste away, and soon die. The first symptom of cholera is a loss of appetite, then follows a jerking or heaving of the sides, which is soon followed by the death of the animal. We are a slipshod set of farmers, and make no investigations for determining the causes of the various diseases which affect our animals. We depend too much on nature, with its sun and rain, and try to go it easy.

Mr. IRA R. FOSTER, Gadsden, Etowah County, Alabama, says:

Horses are afflicted with numerous diseases, the most alarming and fatal of which are "bots" and colic. The former manifests itself suddenly and produces great agony, which frequently results in the death of the animal in a few hours, occasionally in a few minutes, and upon a *post-mortem* examination the coats of the stomach are found partially destroyed by the worms or grub. The symptoms are a disposition to frequently lie down, stretching the head and neck on the ground, drawing up the top lip and showing the teeth plainly, casting the head back behind the fore legs with nose to the body, excessive perspiration, but no swelling of the body. The symptoms of colic are pretty much the same, and the two diseases are often confounded; but in the latter the body is almost invariably more or less distended, and not unfrequently to an alarming extent. We have no reliable remedy for the bots in cases where the animal is violently attacked. The main hope against its deadly ravages is by means of preventives. The colic is more manageable, generally yielding to large doses of carminatives and anti-spasmodics and purgatives combined, such as cloves, pepper, &c., laudanum, paregoric, ether, &c., and salts, castor oil, turpentine, &c. A slug of moistened tobacco inserted in the rectum is worthy of trial. By regular feeding, watering, and exercise the disease would be less frequent. Of horses violently attacked by bots, 50 per cent. die in less than twenty-four hours. Not more than 5 per cent. die of colic.

Distemper, bloody murrain, hollow-horn, and hollow-tail are the diseases which mostly afflict cattle. The first-named manifests itself in and about the head by the issuance of feculent and corroding pus from the nostrils and eyes, with loss of appetite, attended with great lassitude and exhaustion. I have found mercurial purgatives, aided by salts, the most satisfactory remedy. This disease is not so malignant and fatal as in former years. The murrain is common and fatal. It is manifested by a discharge of bloody urine, loss of appetite, constipation of the bowels, fever and thirst, lassitude, and a general drawing up of the body. No favorable remedy has been presented. Cooling cathartics combined with diuretics and diaphoretics have been tried with partial good results. At least 50 per cent. die when violently attacked, and generally within one or two days. Hollow-horn is common, though not necessarily fatal. It shows itself in cold horns and languid looks, loss of appetite, indisposition to move about, seeming great shrinkage in size of body. If neglected, the animal will die from exhaustion in six or eight days. The disease generally gives way after boring with a large gimlet into the center of the horn (which is usually found hollow) and injecting vinegar, table salt, and black pepper daily for several days; also bathing the horn near the head with spirits of turpentine. The hollow-tail is easily detected by manipulating the tail from root to tip. A portion of the bone will be found destroyed by absorption, say from three to ten inches in length. Make an incision to the center of the tail where the bone is missing, and insert a liberal quantity of black pepper and salt. Then close up this orifice and bandage well, and the animal will soon recover.

Cholera is the main disease afflicting hogs. It is common and emphatically fatal, often killing by the hundreds within a few days. On its first appearance it generally selects the best and fattest hogs for its victims. Although many remedies have been tried, and some with apparent success, none seem to be at all reliable. A sure remedy would save millions of dollars annually. A remedy for this disease we need above all others. If found, the farmers could and would advance in prosperity by raising hogs for market as well as for home consumption. If your department can give to the country that remedy, you will have done a great work—one so great, indeed, that its merits and bounds cannot be measured.

Sheep, when in large flocks and closely penned, die by the hundreds of the various diseases which afflict this class of animals. Small flocks in open and broad pastures thrive, and would be remunerative if it were not for the lean and hungry dogs. The rearing of sheep is sadly neglected at the South.

The cholera among chickens is most insidious, and its causes less understood, and it perhaps proves more fatal than all other diseases combined. A great many die with gapes. This disease is caused by nits and mites, which is the result of uncleanly and improperly ventilated quarters. A great many remedies have been tried. A very simple one is to rub the fowl with kerosene oil, and put one or two drops down its throat. This will generally destroy the vermin. The better plan is to keep clean houses, as prevention will be found worth a pound of cure. This is applicable to all classes of farm-animals.

Mr. W. J. EUBANK, Birmingham, Jefferson County, Alabama, says:

There have been no diseases prevalent here among horses since the epizootic influenza, which is still fresh in the minds of the people of the whole country. They occasionally die with colic, inflammation of the lungs, inflammation of the bowels, &c.; but as there are no veterinary surgeons here, little is known of the causes of these diseases. Cattle occasionally die with murrain. Goats are almost free from disease. Occasionally numbers of a flock will die with a malady little known here. They are generally attacked with a fit. When apparently healthy they will sometimes begin turning around, which they will continue until tired out, and then fall down. They may get up soon and stagger about a day or two and die. Sometimes they will lie around three or four days, apparently unable to get on their feet. Now and then one will recover without treatment.

Hogs are afflicted with cholera and quinsy. In the dry weather, during summer and fall, when they are obliged to lie in dust, pigs and young hogs are frequently attacked with a disease that has only been known here some three or four years. Little pimples make their appearance on the body similar to small-pox sores. The skin under the body and inner part of the legs reddens, the nostrils swell, and the patient dies within from three to ten days. I could learn nothing from a *post-mortem* examination. The lungs and intestines appeared natural. The disease is confined solely to pigs and young hogs.

Poultry sometimes have cholera and roup. The former I know nothing of, but the latter frequently occurs in cool, damp weather in spring. The head swells, the nostrils and eyes inflame, and discharge a viscid mucus. The nostrils should be syringed with a solution of carbolic acid or nitrate of silver, and sulphur given internally in feed. Where stock are well cared for and supplied with a variety of food and plenty of salt, they rarely ever suffer from any disease.

Mr. JOHN KENDALL, Amo, Hendricks County, Indiana, says:

The only disease prevailing here among any class of farm-animals is that affecting swine. A diagnosis of the disease, as a rule, seems to be about as follows: First, the existence of a dry cough for weeks before any dangerous symptoms are manifested. Second, refusal to eat, and a disposition of the animal to lie down with its feet under its body. Third, excessive purging in many cases, the excrements frequently being black. Fourth, constipation. In cases where the urine is very yellow, or where bleeding at the nose occurs, death soon follows. Many will linger a long time after they have lost all disposition to eat; others will die within a very few days. The mortality is greatest among pigs. Where older hogs are attacked, from 10 to 25 per cent. recover.

Every hog that dies in this section of country is said to have died of cholera. On examination dead ones were found to contain worms in the intestines. No satisfactory remedy has been found, notwithstanding the many "patents" and "sure cures."

The disease prevails more extensively during July, August, and September, and diminishes as frost and cold weather approach. A lot of my pigs were affected with a cough, as before stated, but about the first of September I had a valuable horse die. I cut the carcass open, salted, and allowed the pigs to devour it. Soon after they commenced feeding on it the cough disappeared, and the pigs have since been apparently healthy. Whether this was due to the fresh meat and change of diet, I cannot say.

Mr. IRA ROWELL, Danvers, McLean County, Illinois, says:

The "hog question" has been discussed for the past ten years in the farmers' club at this place, without any definite conclusion having been reached as to the cause or remedy for the diseases incident to this class of animals. No two persons have ever agreed upon the subject, which has been discussed until it is threadbare. It is now universally believed, however, that alkali in some form is the best preventive of so-called hog-cholera.

Many of those who have taken the best of care of their hogs, and escaped the disease for many years, were at last visited by it, and lost as heavily as those who paid less attention to their stock. At present some of my neighbors are suffering heavy losses among their hogs, while mine are comparatively free from disease, and have been for a number of years. But my turn may come soon. I have a shelter for my hogs, open on two sides, and keep salt and ashes always before them.

Mr. J. T. LAW, Hawk Ridge, Coffee County, Alabama, says:

Cattle in this county have been in good condition for several years past until recently. They are suffering to some extent with a disease called murrain, which proves very fatal. For several years past swine have been afflicted with a disease called cholera. Mine were first attacked during the war, and I have not been successful in raising hogs since. There are various supposed causes of the disease. Some think that an exclusive corn diet will produce the disease. Others think it is brought on by the hogs feeding on mushrooms, which grow plentifully in the bottom-lands. A farmer in Pike County informed me recently that he had been in the habit of giving his hogs nux vomica, and that he had never lost one by cholera. Last spring I commenced feeding this drug in slops from the kitchen, and since then my shoats have been doing very well. I also feed a little corn.

Chickens in some localities are dying rapidly of a disease called cholera. I have heard of no remedy that has a tendency to arrest the malady. The excrements under the roosts are of a deep-green color.

Mr. W. B. DERRICK, Baileyville, Ogle County, Illinois, says:

In regard to diseases of farm-animals in this locality I would say that, during the past year, stock have generally been healthy, excepting that a considerable number of hogs have been affected and some have died. Last summer a disease prevailed among the swine in the western part of this county, and in adjoining counties, which was termed "hog-cholera" by some, but more properly "lung disease," as the attacks were accompanied by cough and congestion, which in many cases soon resulted fatally. On a *post-mortem* examination of some it was found that the lungs were badly diseased, and apparently the direct cause of death.

Within the past few weeks a number of hogs have died quite suddenly in this neighborhood, from a disease supposed to be the veritable hog-cholera, as only certain herds were affected and were soon decimated, while other herds in the vicinity escaped. The hogs affected suffered from purging, and death soon ensued. I am unable at present to give you a definite diagnosis of the disease, nor have I heard of any effectual remedies after the animals have become badly affected. Several preventives and specifics have been tried. The best preventive, I think, consists in keeping the animals in a sound, healthy condition, by feeding them wholesome food and keeping them in clean, well-ventilated quarters.

Last winter a large number of swine in this locality were afflicted with a cough, but the most of them recovered in the spring.

Mr. E. P. CHANDLER, Holly Springs, Dallas County, Arkansas, says:

We have only to note a disease prevailing among hogs, which we term cholera, but it is somewhat different in its diagnosis from the cholera which prevailed to an alarming extent in this section some years ago. The first symptom noticed is an indisposition to eat or take nourishment of any kind, which primary symptom is followed with purging and vomiting in most cases, but in some a complete cessation of the secretions or excretions, accompanied with high febrile symptoms. The duration of the disease is from one to five days. The principal remedies used have been arsenic, turpentine, coal-oil, and opium, and some farmers have even resorted to mercurial preparations, but all with about the same effect; that is, the loss of about 75 per cent. of the hogs that have been attacked. At least 50 per cent. of the hogs in some localities have already died, and the disease still rages with unabated violence.

Mr. EZEKIEL HEMSINGER, Burnt Prairie, White County, Illinois, says:

All the material drawbacks we have here in stock-raising is that among swine, known as "hog-cholera," and from this cause our farmers have, to say the least, been kept down, and some of them have even lost their homes. We have suffered from it now for seventeen or eighteen years, it having reached us in less than twelve months after it first started in Ohio. In the first place, we are convinced that it is a contagious disease, as hogs very rarely take it in any other way than from contact with diseased animals. I live in a hog-raising district, and for twelve years past this has been the universal belief of our farmers. In all this time, with the closest observation, we have not known certainly of a case where hogs were kept in an inside inclosure where others could not reach them.

It is also a well-established fact that hogs have the disease but once. Though some of the herd may sometimes show signs of the disease, they never take it again under any circumstances. A sow may pass through cholera when a pig. If kept for a farrowing sow she will continue to bear her pigs in the midst of a dying herd until she dies of old age, and never again be affected by the disease. What is very strange and unaccountable, is the fact that her pigs, as long as they draw nourishment from the mother, will not take the cholera, but as soon as they are weaned they take it as others do.

The disease usually sweeps over our country once each year. Sometimes two years may intervene, but such a rest we have never had more than once or twice. It generally reappears about eight or ten months from the time of its previous appearance, just as measles and whooping-cough in the human family periodically reappear. We hear of the disease as existing at some distant point, and watch its progress. It gradually approaches until it reaches our next neighbor. If we can now succeed in keeping our hogs and pigs in an inside inclosure, at some distance from the infected ones, they will remain safe; but if they are allowed to smell of a sick hog through the fence they invariably take the disease, which makes its appearance in eight or nine days after being exposed to it.

The first symptom of the disease is a short, quick cough when disturbed, and an inclination to lie in bed. Some will be severely purged and others will vomit, while some will do both. These symptoms are followed by high fever, unusual thirst, and a high, purplish discoloration of the ears, belly, and flank. The duration of the attack greatly varies. Some die within ten minutes after the first decided symptoms manifest themselves, while others may linger a month and then die. The fatality of the disease also varies. Some herds may escape with a loss of 25 per cent., while others may be decimated to the extent of 90 per cent. It is not uncommon to hear of the loss of all in small, well-kept herds. The average loss is about 50 per cent. of all hogs attacked.

As to cures, we have found none. The most successful treatment we have ever found is to keep them away from water and sheltered from snow and rain. It matters not how hot the weather may be, they should have no water either to drink or wallow in. If they have grass or clover to feed on, give them nothing else. It is better for them to have nothing at all for the first week than to feed them on corn. They should not be crowded, and if daily changed from field to field, so much the better.

The majority of writers on hog-cholera seem to know but little about the disease which bears with such crushing weight on this and similarly situated districts. It is claimed by almost all of them that it is the neglect of proper sanitary conditions; but when the disease prevails, it is a well-known fact that among the best-fed and best-grown hogs the fatality is three or four fold that which attends hard-favored, poor shrimps that are but half fed and never properly cared for. We all agree that unhealthy food and foul bedding engenders disease among swine, but that has no relation to our Western hog-cholera.

In all older-settled parts of our country, hogs are restrained from running at large. This is the practice in the prairie counties of Central Illinois, where the disease is not known; but even in this section of the State there are some farmers who shut their hogs up in the barn-lot, where they are compelled to bed in the manure heap, and where they soon sicken and die of filth. Those who raise hogs successfully keep them on clover in summer; and if they have the range of the whole field for choice of bedding and of cover, they will bed in a clean place. We think we have learned by experience that there is no more healthy diet than clover for hogs, yet it is not uncommon for 75 per cent. of those so kept to die in the clover-field.

Some persons urge as an argument against the theory of contagion that the disease must have a start somewhere. We know it has a start; but where and for what purpose, we are ignorant. Isolation sometimes prevents its appearance, but not always. I have practiced this plan, and sometimes have succeeded in preventing the appearance of the disease; but at other times I have failed, and have lost hogs to the amount of $1,000 at one visitation.

Since cholera has proved so fatal among hogs, every sick or dead hog is charged to the account of this disease. Even scientific investigators have greatly erred in mistaking manure-befouled sick hogs for cholera cases.

Mr. A. P. GREEN, Vermontville, Eaton County, Michigan, says:

I have had but few opportunities of making a diagnosis of the diseases which afflict farm-animals. For many years past this section has not been troubled seriously with any contagious disease where fatal results have followed. The epizootic, which traveled over nearly the whole country in 1875-'76, has left many of the horses in this locality in an unhealthy condition, afflicted with discharges from the nostrils, swelling of the glands, and coughing, with rather a heavy appearance of the latter at times, but unlike the heaves. When an abundant draught of cold water is taken the indications of wind-broken breathing cease. The animal becomes quite enfeebled in constitution.

Mr. JOHN G. OXER, Campbellstown, Preble County, Ohio, says:

On the manifestation of the first symptoms of a disease which so seriously and fatally afflicts hogs in this locality, the animal assumes a dull and sleepy appearance, staggers when it attempts to move about, and seems weakest in its hinder parts. It usually wants to hide itself in litter and straw, and when it does so lies flat on its belly. In most cases there are frequent slimy discharges from the bowels, accompanied with profuse bleeding at the nose. Large hogs become very much prostrated within from twenty-four to forty-eight hours, and quite frequently die within that period. The fatality is greatest among pigs from four to seven months old; the disease quite often carrying off from 80 to 90 per cent. of the younger shoats. Among hogs that have attained their growth the loss is from 40 to 50 per cent. The disease is usually attended with high fever and great thirst. The skin is generally covered with small red spots of a very deep color, and before death ensues the breathing often becomes very laborious.

After death the internal structure of the hog is generally found to be perfect, except the lungs. In most cases these important organs are found in a very unhealthy condition, in many cases presenting the appearance of jelly. As regards remedies and preventives, almost everything has been tried, and I can say, from experience and close observation, with very little success. My remedy would be to separate the sick from the well hogs, and kill the infected ones as fast as they show symptoms of the disease.

Mr. J. E. MINTER, Boonville, Owsley County, Kentucky, says:

Hogs in this locality have been afflicted and are now dying of a disease known as cholera. The disease is not so prevalent or fatal, however, as in former seasons. We have no preventive or remedy for the malady. Fowls are also dying quite rapidly of chicken-cholera. When attacked the gills of the fowls turn pale, they lose their flesh, and generally die very suddenly. We have no remedy for the disease.

Mr. J. W. NICHOLSON, Camden, Camden County, New Jersey, says:

Quite a number of horses have died here with something like "dumb colic," a disease which makes very rapid progress; if not relieved, generally ending in the death of the animal within from four to eight hours. The most effectual remedy that has come to my knowledge is sulphate of ether and landanum. Some few cases of "mad staggers" have occurred, for which there seems to be no remedy. I know of no other disease among horses which assumes anything like an epidemic form. We have had a few cases of Texas fever among cattle, for which no successful treatment has been discovered. Cholera among hogs probably kills 5 per cent. of these animals. Losses from chicken-cholera will reach 10 per cent. of all the fowls attacked. No treatment of this disease has proved of any benefit. It sometimes leaves only two or three fowls out of a large flock.

Mr. FRANK HERR, Waterloo, Monroe County, Illinois, says:

For the past year the hogs in my neighborhood have been more or less afflicted with malarial fever. The disease commences with red and sore eyes, which symptom lasts a day or two, when the hog grows stiff, shivers with cold for a few minutes at a time, with intermittent heat as of a fever, after which it dies within from three to four hours. My remedy is twenty-four grains of quinine, given in sugar and sweet milk, in three doses a day; or given in apple-butter forced down the throat, if the animal no longer eats. I found this a pretty effectual remedy. I give pigs three or four grains of quinine per day in some manner.

I have seen a few cases of milk-fever among cows, which generally killed them within from six to fifteen hours. The most lingering case extended into a period of three days. I had good success in administering Glauber's salts and saltpeter in reasonable doses. In the early stages of the attack I gave cold water injections every two hours. If badly constipated I also gave one-half pound of Glauber's salts and a half pint of raw linseed-oil at intervals, until a passage was effected. I then gave two drachms of pulverized camphor in strong valerian tea and kept the cow warm.

Mr. CHRISTIAN HERGENROEDER, Waterloo, Monroe County, Illinois, says:

We have had numerous cases of sick hogs here. The difficulty seemed to be all located in the throat. I gave sulphur in all cases, but it did no good. The hogs would continue to eat heartily up to the time of their death.

Some chickens have died of cholera. I gave berries of bitter-sweet in water, but cannot say that it did any good. The disease continued for about three weeks.

Mr. FRANK ADELSBERGER, Monroe County, Illinois, says:

My personal experience with diseases of animals relates only to hogs. Two years ago, within a period of six months, I lost thirty head. They were a mixture of the Chester-White and Poland-China breeds, and were running in a large dry lot with plenty of fresh water to drink. They were attacked with sore throat, which symptom was soon followed by swelling of the neck. They either could not or did not desire to eat. I gave them lime, coal-stone, and sulphur. The epidemic lasted about six weeks, and the hogs attacked died within from twelve to ninety-six hours after the first symptoms were observed. There was no straw or chaff in and about the lot in which they were confined.

Mr. D. WISHMEYER, Waterloo, Monroe County, Illinois, says:

I can only give my own experience with one class of farm-animals, that of hogs. I had five that were taken sick this fall and but one of them died. Frank Herr cured four of them with a remedy which he gave in sweet milk and sugar. I do not know what remedy he used. The hogs were afflicted with diarrhea and refused to eat.

Cholera carried off a good many of my chickens last spring. I gave them red pepper, but cannot say whether it was of any benefit or not.

Mr. JOHN HERZLER, Huntsville, Madison County, Alabama, says:

In August last a disease made its appearance here among hogs, and by December about all that were affected had died. Up to that time mine had remained comparatively healthy, and none of them had died. I had about one hundred and forty head, and they were running in a plowed field containing about one acre to each hog. I noticed that they kept themselves well rooted into the ground and laid a good deal of the time on their bellies. Before sowing the field to wheat I removed them, and in about a month thereafter they began to die. I lost about all those that had access to the barn-yard and slept in hot places. I penned seventy-five head in a plowed lot containing about one acre of ground, and in March and April, after the lot had become hard and dry, they all died but ten. I think they were affected with typhoid fever and inflammation of the bowels. Some few would become lean and would linger for a long time; but as a general thing they died during the night, although they were apparently in a healthy condition the evening previous. Some few got well. Among those that recovered were some that I fed on warm blood from the slaughter-house. After I turned them out into the woods and swamps they entirely recovered.

Mr. DANIEL GILMAN, Geneseo, Henry County, Illinois, says:

I find that it will be impossible for me to devote the time necessary to make a satisfactory report on so important a subject as that relating to diseases of farm-animals. It is something that ought to be attended to at once in this part of the country, as the hog-cholera sweeps off thousands of dollars' worth of swine every year. I regard the disease as the most important one in this locality, and, from its varied symptoms, I am satisfied it will require a thorough investigation to determine its causes and find a remedy for the scourge.

Mr. JOHN T. GIBBONY, Lamar, Barton County, Missouri, says:

Cattle have suffered considerably in various portions of this county, and quite a number have died from Texas fever, a disease contracted from herds of Texas cattle which were driven through the county. I have heard of no successful remedy for the disease. The contagion was confined to the different localities through which the cattle passed, and did not spread.

Hog-cholera carried off a number of hogs during the past year. Those on the prairies did not suffer to such an extent as they did in other localities. The disease seemed to be more prevalent in the timber-lands and along its margins. Here the hogs were allowed to run at large in great droves. The land was low, and in some places wet, while on the prairies it was dry; besides, they were confined together in small herds.

Mr. W. P. JACK, Russellville, Franklin County, Alabama, says:

Candor compels me to state that as yet I think there is very little real information possessed in this county on the subject of hog-cholera, which appears to be the main disease affecting farm-animals. So far as my information goes, there has been no cure discovered for the disease. It is certain, however, that hogs can be kept healthy by using preventives. In my own experience I find that when I use them I lose no hogs, but if neglected they are apt to sicken and die. The preventives which I have found most effective are such as will keep the lice off them and expel the worms from the intestines. According to my theory they are the main cause of what is known as hog-cholera. I have used tar in early spring, both internally and externally, as a preventive, with unfailing success. Pine seems to be a natural medicine for hogs. In the mountains they hunt for pine roots and eat them freely. Many men who reside in the mountains have told me that they never had a case of hog-cholera, and they attribute the escape of their hogs to the fact of their eating pine roots. Poke-root is another natural medicine for hogs; they root for and eat it freely. It should be boiled with their slop. Sour slop is also a preventive. This should be mixed with charcoal. Frequent salting is indispensable. Copperas is also good as a vermifuge, and bluestone is likewise a fine remedy as a preventive.

An experienced farmer told me that last autumn, after he had lost sixty head of hogs by cholera, he had a very sick one which refused food of any kind. He finally gave it peach-tree leaves, which it ate; he then gave them to the rest of his flock, and did not have another sick hog.

I think your inquiries will do a great deal of good by directing the attention of stockmen to a subject of such vital interest to them. Heretofore, I am sure, those losing hogs by cholera have been too careless in not dissecting them, hence the difficulty at present in getting a correct diagnosis of the disease. Hereafter, should other cases occur, they will be more careful to collect the desired information. I think it will ultimately be found that hog-cholera is not one, but that various diseases are included under this name.

Mr. S. W. COCHRAN, Union, Fulton County, Arkansas, says:

The southern and western portions of this county have been exempt from diseases among stock of all kinds, and stock generally is in fine condition; but the northern and northeastern portions of the county have suffered greatly from a singular and fatal disease among hogs. (I live in the southern portion of the county.) Having heard of the great fatality among this class of animals, and wishing to carry out the request contained in your letter, I saw one of the sufferers in the infected district, and the one I considered most competent to answer your inquiries, and handed him one of your circulars. I sent another one to a prominent breeder in the infected district, and from them I am persuaded you will soon receive full answers to your inquiries.

The gentleman I conversed with told me that at least half the hogs in his neighborhood had died. He said the hogs were differently affected, but death was certain in every case. Being satisfied that those to whom I have intrusted your circulars will write fully in regard to the disease, I leave that to them. But I cannot close without assuring you that I appreciate the object you have in view and the plan you have adopted, believing it may possibly be the means of saving millions of dollars to the farmers of our beloved country.

Mr. A. B. GILBERT, Boonville, Owsley County, Kentucky, says:

There is no prevailing disease among any class of farm-animals in this county except a disease called cholera, which is very destructive to swine. It destroys more hogs than all other diseases combined. No certain remedy is known. We use as a preventive, which is better than a cure, the following prescription, and keep the hogs when sick entirely away from water, viz: One ounce each of brimstone (sulphur), copperas, saltpeter, indigo, borax, and assafetida, well pulverized and mixed with meal or mush; this quantity to be administered to fifty head of hogs.

Chicken-cholera is also very destructive to all kinds of fowls. No remedy is known for this disease.

Some farmers in this county have lost over one hundred head of hogs. Almost all the pigs have died.

Mr. W. B. KENNEDY, Cortland, Trumbull County, Ohio, says:

I have lived as a farmer in this county for sixty-three years, and since the murrain left it, forty years ago, we have had no disease among cattle until last fall, when quite a number of calves, from six to eight months old, died of what we called "blackfoot." It commenced by swelling in the forward legs and shoulders of the animals and affected their breathing. They died within two or three days.

Mr. J. B. RANDALL, McArthur, Vinton County, Ohio, says:

In past years we have had a few cases of cholera among hogs, but have none at present. We think the disease is induced by putting too many of them together, and allowing them to run in the mud and drink impure water. Copperas and sulphur, fed pretty freely, will be found the best remedy. As to cures, all I can say is that we have found none.

Chicken-cholera exists among fowls. Sometimes we succeed in curing them by placing black or red pepper and copperas in their food. We find, however, that those who properly care for their hogs and fowls are never troubled with the cholera.

Mr. ALEXANDER LITTLE, Locksburg, Sevier County, Arkansas, says:

Our greatest losses here are in hogs. A number of remedies are used, the following being the most effective: One teaspoonful each of turpentine, calomel and coal-oil, well mixed and used as a drench three times. As a preventive the following will be found very good: One pound each of copperas and sulphur, and two pounds each of common salt and lye-soap. Mix well with meal or bran and give in slop or dough. I have used this preventive for four years and have not lost a hog, while A. L. Marsh, D. M. Johnson, William S. Ferguson, and many others, have lost hundreds of dollars' worth—two hundred head at least.

Mr. J. M. PETTIGREW, Charleston, Franklin County, Arkansas, says:

There are but two diseases that prevail fatally to any very considerable extent among domestic animals in this county, to wit: The hog and chicken cholera. In this locality the hog-cholera seems to embrace several diseases; and its diagnosis is various. In some instances the hogs have a slow and continuous fever; they become sluggish and seem loath to move; the hair becomes of a reddish color; they have no appetite. In this drooping condition they gradually grow weaker and weaker until they die, but few recovering. In some cases the first symptom is stiffness in the limbs and joints of the hogs; they move as if they were severely foundered. Soon the skin becomes ulcerated over the body, and about the joints and nose boils will break out, emitting an offensive purulent matter. Fever accompanies these symptoms. This type of the disease is very fatal. What few hogs recover shed off most of the hair.

In other instances the lungs and throat seem to be the seat of the disease. The throat and chest become swollen and the animal is afflicted with a cough and difficulty in breathing. These symptoms are attended with fever, and prove fatal in a great majority of cases.

The foregoing are the prevailing types of the disease known here as hog-cholera. By whatever type of the disease the hogs are attacked the same type prevails throughout the entire herd.

No certain remedy has been found. Copperas and blue vitriol are the most successful remedies used here. They are more valuable as a preventive, however, than as a cure. After the hogs have been attacked no remedy has been found to cure any considerable number of them. A variety of food seems among the best of the preventives. During the past summer and fall I fed my hogs copiously on peaches and apples as they fell from the trees, and they have been entirely exempt from cholera and other diseases, while my neighbors' hogs not so fed have died at a fearful rate. Suds from common lye-soap, used for washing purposes, have proved very beneficial in keeping hogs in a healthy condition.

The cholera has killed quite a large per cent. of the hogs in this county during the past summer and fall, and in some neighborhoods it is still prevailing.

Chicken-cholera has also extensively prevailed in this county, and has been quite fatal. The symptoms of attack are drowsiness, the gills and comb become of a purple color, and the evacuations are white and watery. The liver becomes wonderfully enlarged and of a paler color than the liver of a healthy fowl. In most instances the fowl becomes exceedingly fat. Here the disease prevails among chickens, turkeys, and guineas. The most successful treatment for the malady is mercury in some form. I have known that treatment in some instances to prove very successful. The most successful preventives are cleanliness of the hennery, the sprinkling of lime over the floor, and the washing of the walls with lime-water. They should have pure water to drink, in which copperas should occasionally be put. Wild turkeys, even when domesticated, seem exempt from the disease.

Dr. C. M. NORWOOD, Bluff City, Nevada County, Arkansas, says:

All animals, except hogs, have been remarkably healthy for several years past in this section of country. We have had a disease prevailing among swine which has proved very fatal to nineteen-twentieths of those that have been affected. The disease has been called "hog-cholera" among farmers; but from observation and some investigation I am led to conclude that cholera is a misnomer. From the most prominent symptoms I consider it to be a lung disease altogether. The symptoms are, first, great depression, followed by languor and indisposition to move about for the first four or five days. Second, a slight, dry cough, attended with intense febrile excitement and dryness of skin. At this stage there is complete loss of appetite, and *crepitus* is audible in the thoracic region. In this form of the disease death ensues about the ninth day. *Post-mortem* investigation reveals the stomach, bowels, liver, spleen, and pancreas healthy, but the lung hepatized, the air-vesicles filled with sanguino-purulent infiltration from the cellular tissue of the lungs, revealing the fact clearly that there has been great and destructive inflammation of the lungs. We must, therefore, conclude from the symptoms and pathological anatomy revealed by this examination that it is *pneumonitis* of an acute form. We have noticed some hogs that ate heartily and appeared perfectly healthy in the evening, and the next morning were found dead. On *post-mortem* examination this class of animals revealed congestion of the lungs, extravasation of blood into the air-vesicles to so great an extent as to lessen the caliber by infiltration, producing death by asphyxia or strangulation. I consider this the most violent and pernicious form of this lung disease.

Another class of subjects are those that recover finally. I consider this to be the acute form, terminating in a typhoid form. The duration of this type of the disease is from about ninety to one hundred and fifty days. Generally, when the disease assumes the typhoid form, there is some purging from the bowels, and this symptom, I presume, has led many to give it the name of "cholera." I consider it altogether a lung disorder, as it presents itself in this locality, and a proper study of the disease would no doubt convince many that they are laboring in error in their diagnosis of this fatal and malignant malady.

As to treatment, none has ever been adopted that has proven satisfactory. A multiplicity of remedies have been used by the farmers, but all have signally failed. The only remedy I can give that I consider at all reliable is twenty grains of calomel and one and one-half grains of tartar-emetic mixed and given every other day during the febrile excitement. After the fever has subsided give nourishment freely, such as slop from the kitchen, cooked vegetables from the garden, mush (corn), &c.

As to the prophylactic treatment, I know of none. I think the poison producing the disease floats in the atmosphere, and that it is not produced from any local cause. The best preventive that presents itself to my mind is to move the herd to some thick forest as soon as the first symptoms of the disease are observed, and not allow them to run in fields or around the farm.

I hope this short and imperfectly-written note may lead some mind to a more thorough investigation of this important subject.

Mr. W. B. SHAW, Beverly, Washington County, Ohio, says:

Lambs in this locality have been scourged for several years past with a disease called "paperskin," which seems to be worse in wet than in dry seasons. It is not uncommon to lose an entire flock by the disease. It attacks the lambs at the age of from three to five months, and those in good flesh are as liable to it as those that are in poor condition. When attacked, they become very pale and weak, apparently almost entirely bloodless. The stomach contains small red worms, and frequently, in addition, the animal will be found to have tape-worm. I know of no cause or positive cure for the disease. I have tried many remedies, and have found more benefit from feeding pumpkins than from anything else.

Many sheep die with grub in the head. The symptoms are bloody, mattery discharges from the nostrils. Pine-tar placed in their salt-troughs from June until September (during the season the gad-fly deposits its eggs) will be found a preventive. A positive cure will be found in syringing the nostrils with a decoction made from tobacco.

Chicken-cholera is very common here. We know of no cause, nor have we a remedy, for the disease.

Mr. D. STICKEL, Monticello, Pratt County, Illinois, says:

There are no diseases prevalent among farm-animals in our county except among hogs, and this class of animals has suffered more severely this fall than for many years past. Many persons have lost entire herds. The various symptoms of the disease are as follows: The hair inclines to stand erect; a hacking cough; standing around with the nose to the ground; sometimes they have the thumps; frequently they bleed at the nose; some are affected in the head, the eyes matter and frequently burst; sometimes the tops of the ears get raw and are covered with clotted blood; sometimes they are purged and at other times they are constipated.

The duration of the attack varies considerably. This fall the duration of the disease seems shorter than usual, the animals generally dying within from one day to a week after the first symptoms are observed. Once in a while they will linger for weeks, and then die apparently like a person afflicted with consumption or typhoid fever. I think the fatality is nine-tenths of all attacked where no remedies are used. Quite a number of remedies have been used at different times, but with little effect. Sometimes a remedy will appear to be quite successful for a time, but will finally seem to lose all virtue as such. This is especially the case with May-apple root. A Mr. Combs, of our place, has prepared a remedy that is being used considerably in this section of the country, and it appears to have some merit as a curative. He regards it as a certain preventive. He has now used it about seven years, and says that he has never had a case of cholera among his hogs since he has been using it as a preventive.

Mr. J. WESTLAKE, Troy, Miami County, Ohio, says:

Hog-cholera prevails to a considerable extent here and is quite fatal. The epizootic has also prevailed to some extent among the horses of this county, but has not been very fatal. Chicken-cholera prevails extensively in some neighborhoods, and is very fatal.

Mr. R. J. WILLOUGHBY, Federalsburg, Caroline County, Maryland, says:

We have a disease among fowls here which seems to affect but two classes, viz., turkeys and barn-yard fowls. The disease is generally known as cholera. It is very fatal, and kills entire flocks sometimes within the short space of twenty-four hours. It seems to strike in spots. For instance, while the flock of one farmer may be entirely decimated, that of another, who may not reside three hundred yards away, may entirely escape. We have not been able to find any remedy for the disease.
A number of horses were lost during the past summer and fall by farmers in the adjoining county of Dorchester, by a disease known as blind-staggers. A remedy for the disease was extensively used and proved quite successful. It was, to split the horse's forehead and bind horseradish in the cut. In every case where this remedy was used in the early stages of the disease the animals recovered. From sixty to eighty horses died of the disease in that county.

Mr. E. ARCHER, Lancaster, Franklin County, Ohio, says:

The only disease affecting farm-animals here is cholera among swine, and its symptoms are as varied as its treatment. The duration of the disease is from four hours to as many weeks. Nine cases out of ten prove fatal. Our experience here is that there is no remedy for the disease, but we have a pretty certain preventive, viz: Salt, copperas, and wood-ashes, in the proportion of one pint of salt, one-fourth pint of pulverized copperas, and three gallons of wood-ashes, well mixed, and placed in dry sheds, where the hogs can have access to it at all times. This is the only satisfactory preventive to my knowledge. When the disease has progressed so far as to cause the loss of appetite, I regard it as the next thing to incurable. When a hog once refuses to eat, he is dead, or might as well be.

Mr. JOHN L. S. DEBAULT, La Rose, Marshall County, Illinois, says:

In my county diseases are prevailing among hogs to a very alarming extent. Different lots seem to be differently affected. Some have symptoms of quinsy, while others seem to be afflicted with the old cholera, a disease not very prevalent this fall. However, almost every ailment among hogs is called cholera. An entirely new phase of the disease seems to be prevailing this season among my own hogs. They had the run of a very large pasture, comprising creek-bottom and upland, with an abundance of young timber. They had pure running water, a fine blue-grass pasture, an occasional feed of corn, and in addition followed a herd of corn-fed steers. I had two hundred and thirty-three head, and I thought they were the finest lot of shoats I had ever seen—healthy in every respect apparently, and thrifty. October was very warm until toward the close of the month, when we had a sudden change to severe cold weather. My hogs were at once affected. They commenced to sneeze and cough, and the pupil of the eye turned white, causing total blindness in a few hours. Death would generally ensue within from ten to twenty-four hours. Their bowels did not seem to be affected; the disease seemed to be entirely located in the head and nasal organs until within two or three hours before death, after which the whole trouble appeared to be with the lungs. I think the symptoms were those of catarrh. I tried various remedies without any good effect. Among other things I did was bleeding, but this only seemed to hasten death. I then tried turpentine, sulphur, and copperas with like ill success. Finally I sent to Grundy County for a hog-doctor, who had great success in killing all he undertook to cure. I changed the quarters of those that remained, placing them in dry hospital buildings, in small lots together, where I could give them medicines at pleasure. This did not stay the disease, as the confinement appeared to cause it to rage with greater virulence than before. I finally lost two-thirds of my herd—one hundred and fifty-five out of two hundred and thirty-three—before the disease abated. My opinion is that the disease was caused by too high a temperature of the body when the sudden change of weather took place in October, and the consequent sudden cooling of the outside surface.

Mr. John Frost, Hoboken, Hudson County, New Jersey, says:

Our horses have suffered greatly by epizootic, which seems to have been chronic, for the last three years. The symptoms are as follows: The eyes become dull and heavy, the glands of the throat swollen, loss of appetite, followed by a copious discharge of mucus from the nostrils. My system of treatment was as follows: I had my stable thoroughly cleaned, and gave it several good coats of whitewash prepared from ordinary lime. I then fumigated it once a day by burning pine-tar, being very careful to close the door and keep all the smoke possible from escaping. About noon I would prepare a feed for them by scalding about three quarts of wheat-bran, and after adding about one gill of cider vinegar would feed it to them warm in a nose-bag. If they refused to eat they at least inhaled the steam from the food. This treatment seemed to bring them back to their appetites. I fed them young carrot-tops, which they devoured with avidity. At the end of four or five days with this treatment the horses were ready to go to work again. Some of my neighbors refused to follow my treatment and called in veterinary surgeons, who were in most cases from four to five weeks in getting the horses on their feet again. In a great number of cases very valuable animals were lost, while my own thrived and recovered their wonted spirits and strength in most cases in less than a week.

Horses in this district suffer greatly from inflammation of the bladder, brought on in most cases from fast driving or heavy pulling. The symptoms that have come under my notice are as follows: The horse frequently stretches and attempts to stale, but cannot. I have tried niter and gin, in fact all ordinary prescriptions given by veterinary surgeons. They failed, and I resorted to my own treatment, which is as follows: Take about twenty-five or thirty roots of parsley, stew them in about three quarts of water, strain them through a collander, and give the horse as a drink one pint every half-hour. The second or third dose has never, in my experience, failed to relieve them.

Dr. Frank Prince, Jonesborough, Jefferson County, Alabama, says :

There is a disease prevalent here among hogs which for years has been known as cholera, but which should more properly be termed measles. The first symptom that manifests itself, on close scrutiny, is seen in the hog walking on its toes, and not upon the entire foot. But for some time previous to this the hog has been affected, and this is the result of contraction of the intercostal and abdominal muscles. There exists a latent inflammation of the parenchyma of the lungs, and cutaneous or superficial fascia, which causes the hog to contract the muscles for relief, hence he pitches on his toes. He has been having fevers several days, as is manifest by dullness and stupidity, indisposition to play, the head bowed with the nose close to the ground, and a thin, viscid mucus dropping from the mouth. Now examined, the mouth will be found inflamed, an eruption is visible in and around the throat, and the appetite is fast failing. A slight cough has set in, accompanied with occasional vomiting. The eruption soon fastens itself upon the entire alimentary tract, so that the stools soon become thin, purulent, and bloody. Great emaciation supervenes, and the hog staggers in walking. Purulent matter and blood are sometimes passed off by the animal. The hair begins to fall off as the hog becomes more and more emaciated, and a small miliary eruption is to be seen all over the skin. Without relief he will soon die. Sometimes he dies much earlier in the attack, which is caused by this purulent matter entering the blood, by which means it is conveyed to the heart and brain, and causes the animal to turn round in a circle until it drops dead. Could this eruption be thrown out at the commencement of the attack, and the hog kept for one week in a dry house where there is no dust, he would soon recuperate. But where measles is complicated with an inflammation of the bowels or lungs, with the usual exposure to which all hogs are subject, death is almost inevitable. Hogs that are taken up and put early on treatment are apt to recover, or at least the mortality is not so great.

There are almost as many ways for the treatment of this disease as there are sections of country in which it occurs. One old and successful farmer told me that he always kept slops for his hogs made of corn or meal boiled with ashes or poke-root, and that he rarely if ever lost a hog. Another stated that he used ashes, salt, copperas, and sulphur with great success. The great secret in all this treatment is the alkali that is used. When this is administered in time it acts as an alterative, controls the secretions of the mucous coat of the intestines, stimulates the absorbents, sets up a healthy action in the lymphatics, causes the skin to assume a healthy function or action, and the disease soon disappears. So you see every one has his remedy so convenient that there is no necessity of going from home to obtain it. It consists in the proper use of good wood-ashes and salt.

Mr. L. G. Maynard, Hampden, Geauga County, Ohio, says:

With very few exceptions no general diseases have prevailed among farm-animals in this county since the "bloody murrain" left us forty or fifty years ago. The epizootic prevailed among our horses a few years ago, but comparatively few cases proved fatal. At that time warm stabling, light food, and exercise proved to be the most efficient remedy. For a few years past I think heaves and other lung complaints have prevailed to a greater extent than fifteen years ago. A little lobelia, say a teaspoonful, once in two or three days, together with straw and provender (oats and corn ground), seems to be the most approved remedy.

I lost a valuable mare last spring which appeared to have every symptom of pulmonary consumption. She began with a slight hacking cough, which increased steadily until April, when she died very much emaciated. No remedy seemed to even check the disease or afford temporary relief. Such cases, however, are rare.

It is very rarely that meat cattle are either affected or die of disease here. On Thursday I met with our Farmer's Club, and laid the subject of your circular before it. The testimony of those present generally corresponded with what I have above stated. A few instances were reported of cattle having what some called "dry murrain," or food drying up in the first stomach. The following remedies have been used for this trouble with success by different farmers: Linseed-oil; one pint of flax-seed boiled with three or four quarts of water; saleratus and buttermilk; spirits of turpentine.

A few flocks of sheep are affected with "foot-rot" or "hoof-ail." Remedies are used with success, which makes the damage to flocks comparatively small.

The only cases of hog-cholera I have ever heard of were three or four years ago, and occurred in a drove from the West, which were peddled out to the farmers and factories in this locality. They evidently brought the disease with them. I heard of one man who lost eleven head, and another one who lost two or three. No remedy was of any benefit. The soil and climate of this county seem adapted to the healthfulness of hogs; but too little corn is raised to make hog-raising for market profitable.

Dr. J. M. Johnson, Locksburg, Sevier County, Arkansas, says:

As a physician I have been engaged in the practice of medicine in all its branches for the last twenty years. I have also had a farm, and have given a good deal of attention to stock-raising upon a small scale. As to the names given to the diseases affecting our farm-animals, they are generally so far established that, whether suitable or not, it would be hard to change and eradicate them from the minds of the people. Horses, cattle, and sheep here, according to my observation, are comparatively healthy, although, like all mortal creatures, they are subject to disease and premature death. For an animal occasionally to become diseased, sicken, and die is something we naturally expect; but what alarms us most are the destructive epidemics which, for the past twenty years, are existing somewhere at all times, killing our useful and indispensable animals, as well as our much-relished and profitable fowls. Hogs and poultry here seem to suffer most from the ravages of disease.

Hog-cholera, meningitis or staggers, quinsy, and mange are by far the most common diseases among swine. The symptoms of cholera are: The hog is obviously sick, mopes about and lies down most of the time, occasionally vomits or tries to do so, eats but little or none at all. In a day or two it will perhaps have superadded a profuse diarrhœa. If the disease runs a regular course the animal will continue to vomit and purge until the alimentary canal is emptied of all its feculent or substantial contents, followed by watery or serous and sometimes bloody operations, with cramping of the muscles and particularly of the bowels. When all the above-described symptoms are seen the complaint has reached its second stage, and is in its height or at its acme of apparent force. Here, if it does not yield to the efforts of nature with the aid of remedies, the hog will pass into the last or declining stage. If the disease yields, the animal will continue warm and all the symptoms will begin to moderate. If not, it will go into collapse, become cold, or nearly so, continue to strain and cramp and utter low grunts, and sometimes will even shriek with pain. The duration of the disease is a good deal owing to its severity. Generally it lasts from one to four days. All cases that result in death do not run the same course. Sometimes all of the above symptoms are not present. Some epidemic symptoms are milder than others, but all seem to be malignant, for nearly all the hogs die that take it if left alone. The same epidemic is not equally severe in different cases. Sometimes the attack is so violent that the animal is in the last stage from the outset, or it may die from nervous prostration with no reaction, vomiting or purging.

The diagnosis can easily be determined by the symptoms when they are all present, especially if the hogs are in living order, and the weather is warm; for, according to my observation, the disease prevails almost entirely during the summer months. Of the causes of the disease I can say but little, because they are not perfectly known; but we know that hog-cholera is epidemic, and that it is a poison, very irritating in its action upon the stomach and bowels; that it has a preference for localities, and prevails more generally upon the borders and in low bottoms than upon lands that have been previously overflowed. That it is also contagious we have some good reasons to believe. One thing I do positively know—that there are some powerful predisposing causes that can, I believe, be almost or entirely prevented. I will leave this point for more time and evidence, as I can only hint at the subject generally at present.

Fortunately this disease, though very fatal and destructive, often readily yields to proper treatment when administered in time. (By far the best plan is the preventive treatment, which is comparatively cheap.) The following prescription will be found valuable: One quart pure alcoholic tincture of camphor, one-fourth pound each of prepared chalk and *Hydrastis canadensis*, one pint of tincture of catechu, and one-half pint of laudanum. To administer this prescription lay the hog on its back, place a stick transversely between the jaw-teeth, and pour down one ounce of the mixture once every two or three hours. If the first and second doses do no good, it is almost needless to persevere. The mixture should be well shaken before using. There may be other indications that could be met by proper medicines, but generally if the above fails we may as well let the hogs go.

If we carefully examine a hog that has died of cholera we will find the liver and kidneys diseased. The coatings of the stomach and bowels will also be found more or less inflamed from great irritation. We may also find patches of ulceration, with worms imbedded about the kidneys and mesenteric glands. During the prevalence of epidemics some hogs may escape the disease, while others may have it in a mild form.

Some years ago I saw a preventive advertised in a Tennessee paper, which I adopted in part, as there were some incompatibles in it, and I have found it a complete preventive not only of cholera, but of all other diseases affecting swine. It acts gently and mildly on the liver and keeps it healthy; in a word, it is tonic, diuretic, alterative, and anthelmintic in its action. It is composed of the following ingredients: To one gallon of tar add four ounces of calomel, one-half pound of copperas, and one-half pound of golden seal. Stir the ingredients well, and with a wooden paddle spread it lightly upon an ear of corn, and give one ear to each hog or shoat once every three weeks. When diseases are prevailing extensively give one prepared ear every week. When hogs are hungry they will eat every grain of the corn and will finally seem to relish it.

In answer to the question as to the average fatality from diseases among swine in Arkansas, I believe over half of the number die before they are ready for slaughtering. There are a great many things recommended as preventives and remedies which I have no confidence in whatever.

Mr. J. J. Litton, Alton, Oregon County, Missouri, says:

With the exception of hogs, all classes of farm-animals in this particular locality have been extremely healthy for some years past. For six months past a disease generally known as cholera has been working sad havoc among hogs. But few large animals have died from the disease, but a great number of pigs and shoats have been lost. The first indications of the disease are seen in the animal becoming stupid, in which condition it continues until relieved by death, which occurs within from one to four days. Sometimes the throat appears to be affected, and in many instances the feet swell and burst open.

Mr. G. W. Cullison, Allerton, Wayne County, Iowa, says:

The worst disease among hogs that I have noticed within the last twelve months has the following symptoms: 1. An indisposition to eat, accompanied with drowsiness. 2. Vomiting occasionally. 3. The skin becomes cracked and sore, with increased vomiting and an indication to thump. 4. Thumping increases in severity; skin rolls in folds. 5. Diarrhea sets in, and this and thumping close the scene.

The disease seems to run in families, but is not otherwise contagious. The mortality reaches from 30 to 50 per cent. in a family. No remedy has been discovered by me, but with cleanliness and variety of food the percentage of mortality may be much diminished.

No name has been given the disease, but many call it cholera. It assumes its worst forms during the hot months, especially if hogs are kept in dry pens with no grass and but little shade.

Mr. C. B. Combs, Lamar, Barton County, Missouri, says:

Hogs in this locality have been more seriously affected by disease than any other class of farm-stock. The disease is supposed to be cholera, and the losses have been quite numerous. All kinds of remedies have been tried, but nothing has been discovered that proves of much benefit. There have been some losses among cattle by what is called by some Texas fever and by others dry murrain. The only remedy that has proved of any value is a purgation of some kind. The animals should be taken from the range they have been accustomed to and put up in close pens and fed green fodder, which has a tendency to keep them well scoured out.

Mr. WILLIAM JOHNSON, Saville, Crenshaw County, Alabama, says:

The first cases of hog-cholera that came to my knowledge were in East Tennessee in 1863 and next in Alabama in 1874. The disease appeared to travel south. The best remedy I ever tried was strong lye in food or the tea of poke-root mixed with corn-meal; also a mild tea made of May-apple or mandrake mixed with meal appeared to be more effectual than any other remedy tried.

Mr. JAMES WEILER, Alburtis, Lehigh County, Pennsylvania, says:

I give you the following preventive prescription for hog-cholera: Salt the hogs twice a week regularly, using a teaspoonful of pulverized copperas to every four quarts of salt; or, for 100 hogs, procure 50 bushels of clean wood-ashes and mix therewith salt and sulphur and ten pounds of pulverized mandrake-root, and scatter one-half the amount where the hogs can get at it; and at another time scatter the balance in the same way. The pulverized mandrake-root acts vigorously on the liver, and both sulphur and ashes are good remedial agents in common use. Or take two parts of sulphur, one part of antimony, one part of saltpeter, one part of copperas, and a small portion of asafetida, mix with salt and place in a trough in a dry place where the hogs can at all times have free access to it.

As a preventive for diseases of chickens use asafetida and some finely-ground black pepper. Put the same in a piece of cloth or rag, nail it in the bottom of the vessel where you water your chickens, or, if the vessel is iron or stone, lay it on the bottom and confine it there. You should not use too much at one time. One-half ounce every two weeks will answer for fifty or sixty chickens.

Mr. EVAN GOOD, New Vienna, Highland County, Ohio, says:

Hog-cholera, so called, is and has been for three years alarmingly prevalent in this and adjoining counties. Last year the disease was terribly fatal, probably from 60 to 70 per cent. of those attacked dying. This year the fatality has not been nearly so great, simply for want of material, farmers in nearly every instance having sought and found a market on its first appearance in their herds. An exception should be made of those cases where the whole herd was attacked at or nearly at the same time.

You ask for a complete diagnosis of the disease. That would be a task to appal the stoutest. Probably no two men could give the same report. Scarcely two animals on the same farm are held in the same way. Pigs from six to ten months old die faster than those of twelve months and upward. Many more die after the fattening season commences in the fall than at any other time. Hogs having a wide range of woods pasture are less liable to infection and more likely to recover when attacked than those confined in pens or small lots, notwithstanding the danger of contamination would seem to be greater.

Here are a few of the symptoms as I have seen them and as they have been reported to me by neighbors: Fever in nearly all cases; a dry cough is often a premonitory symptom; vomiting; purging; bleeding at the nose; bleeding through the pores, particularly about the head; paralysis of the hinder parts; giving way of the fore legs; dropping off of the ears and tail; constipation. One man who saved eleven head out of one hundred and four this fall says that the lungs, or the portion next the heart, was always diseased, while a membrane which surrounds the heart was filled with water. Sometimes, while eating, a hog will give a squeal of agony, jump a foot or two from the ground, and fall dead. No cure has been found. Turpentine and capsicum are the only preventives I know of that are worthy the name. They will not always prevent, but they have the effect to brace the system for the attack. It is the opinion here that those who have seen the most of the disease know the least about it. Those who have not seen it have at least a theory. Those who have suffered by it come out of the siege with their theories crushed. It is the most confounding and bewildering disease that can be imagined; it will not be investigated. Let the department dig this thing up and it will have the everlasting thanks of this plague-ridden section. But do not let the investigator enter the field with a theory or he will be disgusted at the outset. Let him follow facts and base his theory thereon.

Mr. GEORGE T. McWHORTER, Chickasaw, Colbert County, Alabama, says:

I send you, in alcohol, by to-day's mail a number of worms taken from the lungs and intestines of hogs that died during the epidemic last summer. This disease was called cholera by farmers in this vicinity—a term, by the way, which is here used to cover "all the ills that hogs are heir to." These worms are from two different hogs, several miles apart, and show the identity of the trouble. The small worms are from the alimentary canal; the larger ones mostly from the lungs, although nearly all the tissues were to some extent infested with them. I found the bowels constipated, notwithstanding the name applied to the disease, and filled with impacted fæces. Mixed with the fecal matter and adhering to the walls of the canal were myriads of the small worms. I saw no large worms in the bowels. There were numbers of small inflamed points along the inner surface of the bowels, but no large patches of inflammation. Perforations were perhaps made at these points.

The hogs had been troubled with persistent cough, which led me to examine the lungs carefully. Here were found great patches of inflammation, and the larger worms were very numerous. The lung tissue in places was entirely broken down and the sounder portions riddled with worms. Next to the lungs and bowels the liver suffered most. The worms here were also larger than those in the bowels, from which I infer that the worm, after being hatched in the bowel, perforates it and penetrates the other tissues. Some fattening hogs recently killed show worms in the liver; but as the hogs seemed tolerably healthy, with sound lungs, I doubt their identity with the ones sent you.

From what I have seen of the disease I make the following deductions:
1. The worms are hatched in the bowels.
2. They must be destroyed before they leave the bowels.
3. When the lung is perforated treatment is unavailing.
4. Almost all cases let alone prove fatal.

Treatment should be founded on these principles. I recommended calomel and arsenic to a number of farmers. Many hogs just taken recovered under this treatment, but nearly all the old cases died.

Mr. SAMUEL BARR, Amanda, Fairfield County, Ohio, says:

There have been no contagious diseases prevailing among farm-animals in this neighborhood except a disease known as cholera among hogs. It has been very fatal in this vicinity, and several hundred hogs have recently died from its attacks. The disease does not attack all alike. Some commence by bleeding at the nose, others by vomiting and purging. Still others will quit eating, lie around a few days, and then die. Some will eat with apparent good appetite and in an hour will be dead; some will linger two weeks and then die. About 5 per cent. of those attacked recover. No reliable remedy has as yet been discovered. What seems to benefit one herd has no effect on another. It is believed by many stock-raisers, however, that the disease can in a great measure be prevented by using such remedies as are within the reach of all. Give your hogs comfortable quarters and plenty of good water. Salt them every week, and mix with the salt wood-ashes and sulphur; have stone-coal for them to run to, and feed regularly. Those who have practiced this treatment have saved their hogs.

Mr. J. K. PRUDEN, Sidney, Shelby County, Ohio, says:

I have had a great many hogs to die of a disease called cholera. When it first made its appearance on our farm it was very fatal. The animals were handled in various ways. Some would vomit, some would purge, some would do both, and some would do neither. The few that recovered would break out all over and lose their hair, and in some cases the hide with it. In some instances the flesh would slough off in large lumps. A few of such cases recovered and afterward made fine hogs. I think the disease is a brain disorder and an affection of the lungs, for the hogs have a cough, and the vomiting and purging are no doubt the result of deathly sickness.

I have tried every remedy I could hear of without any success. What would seem to benefit one hog would do no good in other cases. The best thing I have found is sulphur and asafetida; they, however, seldom cure the disease, but they are good as preventives.

For two years cholera prevailed extensively in our flock of chickens, and we lost a great many. Finally, we cleaned and limed their roosts, put in plenty of ashes for them to wallow in, and gave them milk to drink, since which time we have lost none.

There is a disease among sheep here called "sore mouth," which, if let alone, proves very fatal. A preparation made of vitriol and chlorate of lime, and used as a wash for the mouth, will be found a sure cure. I have seen cases of the foot-rot and scab, and I believe the disease the result of too close shedding. I generally keep from one to five hundred head of sheep on my farm, but I am not troubled with this disease. I have sheds for my sheep, but I do not confine them.

Mr. A. H. WRENN, Mount Gilead, Morrow County, Ohio, says:

There has been a slight return of the epizootic among horses this fall, accompanied with a slight cough and a little discharge from the nose. But little medicine was given. Bran mashes and other laxatives to keep the bowels open, with a little extra care, have restored them to ordinary health.

Sheep are affected with foot-rot, scab, and what is known as grub in the head. A good many remedies are used, sometimes with success and again without any apparent effect.

We sometimes hear of a few cases of thumps and cough among hogs, and now and then a case of blind staggers, but few deaths are reported. Charcoal, ashes, salt, and even soft-soap, are used as remedies, especially when cholera prevails among hogs.

Thousands of chickens die annually from diseases incident to fowls. Many families lose large flocks entire. Wild-cherry and white-oak bark, dog fennel, and red and black pepper are used as preventives and remedies. The most successful treatment of late is a small quantity of assafetida in water, blue mass in very small pills, and a little blue ointment on the head.

Mr. JAMES M. BURT, West La Fayette, Coshocton County, Ohio, says :

I can say that, during a residence of near half a century as a farmer in this county, with few exceptions the cause of disease among and loss of farm-animals has been the result of neglect and improper treatment. Notwithstanding the best of treatment, however, the epizootic prevailed for a time among horses ; and what is known here as "colt distemper" frequently prevails, which, if not properly treated, terminates in glanders, an incurable disease. My treatment, which proved effectual, was saltpeter dissolved in hot water, mixed with wheat-bran mash and fed warm with oats or chop feed—one ounce per dose every third day.

No contagious or fatal disease has prevailed among cattle in this vicinity. Feeding at regular hours in winter, with free access to water and salt at all seasons, has been my system, and I have lost none from disease.

Grub in the head has prevailed among sheep. The disease is incurable, but it may be effectually prevented by giving them salt mixed with dry wheat bran as often as once a week during the summer and fall months, when the fly abounds which causes the disease. The same treatment will prevent the disease commonly called " rot," or cure a cold contracted by exposure or sudden changes of the weather. I am not familiar with the foot-rot or scab, as it has not appeared in this vicinity.

The cholera, kidney-worm, and other diseases that hogs are liable to in some localities are effectually prevented by giving them free access to the slack or waste from our bituminous-coal mines, which abound in this vicinity. Copperas and sulphur are its component parts. I have never lost a hog from disease.

In-and-in breeding is believed to be the cause of all the diseases that fowls are liable to. Since we have annually marketed or exchanged all our own raising of males and kept our hennery cleanly whitewashed and the floors covered with lime, we have lost no chicks or grown fowls from cholera or any other disease.

Mr. J. S. ELDER, Darlington, Beaver County, Pennsylvania, says :

Sheep are the only class of farm-animals subject to any specific disease, and the most troublesome one is that known as " pales." The remedies are turpentine and copperas mixed with salt and placed in boxes in their feeding places. But I find they never recover their former health. They dwindle away for a year or two and then die. I find it almost useless to try to save them. Foot-rot also prevails to some extent among sheep. About the only remedy used is sulphate of copper.

Two horses died in this neighborhood a few days ago. They were sick but a few hours, and during this time walked around with their heads down and ears drooped until they fell down dead. We have no veterinary surgeon in this vicinity, and therefore I can furnish you with no diagnosis of the disease.

A great many cows annually die here with puerperal fever. There seems to be no remedy for the disorder.

Mr. RICHARD WRAY, Richmond, McHenry County, Illinois, says :

I have been breeding stock in a small way for forty years, and during that period have had diseases among my hogs three or four times, but fortunately they did not do much damage. Four years ago my hogs showed symptoms of disease. When first discovered two of them could not walk. The place where they slept I found to be damp and wet. Several of the animals were stiff in their joints, and in addition were coughing. I bled the two that were the most seriously affected, in the mouth, and put them in a hole in the horse-manure pile. I covered them all over, with the exception of the nose, with the hottest manure in the heap and then poured two bucketfuls of cold water over them. This was in the morning, and I left them there to steam until night. I then took them out and they appeared to be well. I had to leave home that day, and was absent for several days thereafter. I ordered my men to do the same thing with other hogs that might be similarly attacked. The next day two other hogs were taken sick in the same way, but instead of putting them in a hot place in the manure they put them where the temperature was very low, and the result was that they both died. By separating them into small lots and giving them dry beds to sleep in we lost no more.

I have had sucking pigs affected with fever, hard breathing, and costiveness, for the removal of which difficulty I have used a syringe with some success. The second stage of this disease is hemorrhage of the bowels, of which the pigs die. The principal cause of the disease, I think, is the lack of dry, warm beds, and the sleeping of too many together.

Several farmers near me have sustained severe losses among their hogs this year. One of them told me that spirits of turpentine had been of more benefit to his animals than anything he had tried.

Chicken-cholera has prevailed to a considerable extent. In many cases the disease has proved very destructive. It commenced in my flock by attacking two turkeys which I had bought. I discovered it by a peculiar chirping noise which they made. On examining their mouths and throats I found them almost closed with a fungous growth. This I scraped off, and then applied quick-lime liberally. Two applications of this entirely cured them. I also sprinkled lime plentifully about the roosts where the fowls could get it easily.

Among young cattle we frequently have black-leg, and milk-fever sometimes prevails among cows. Abortion among cows also frequently occurs and often proves fatal.

Mr. ORLANDO WILCOX, Hinckley, Medina County, Ohio, says:

Some time last summer Mr. Whipp, of this county, went to East Saint Louis and bought ninety head of what are called Cherokee cattle, but their long horns, long legs, and gaunt bodies indicated plainly that they were of Texas origin. He brought them to Berea by rail, drove them home, and put them on to what is known as the Wilson farm. Some time in the early part of September his native cattle began to die, and kept dying until he lost about thirty head. I ought to have said before that these Cherokee cattle were very unruly and went almost anywhere they desired. They jumped into most of the neighboring farms, but were driven out as soon as discovered. Among other farms they trespassed upon was that of Lewis Conant, where they were not discovered until they had lain down to rest. Soon after three of Mr. Conant's cows sickened and died. Upon close investigation it was discovered that these Cherokee cattle were infested with wood-ticks, which it was supposed they brought with them, as ticks are scarce in this country. The theory of the farmers in this township is that the Cherokee cattle communicated the ticks to the native stock. Our native cattle, not being used to them, had their blood poisoned, while the others, being used to them all their lives, were not affected. People going from a healthy country into a malarious district will have the fever and ague and other bilious complaints, while the natives who have always lived in that locality are but seldom attacked.

The cause of the Texas fever was much discussed in our local papers. Many contended that it was caused by saliva left on the grass by the Texas cattle; others that it was caused exclusively by the ticks brought by the Texas cattle. It is certain that Whipp's cattle were badly infested with ticks, as a large number were collected and sent to the editor of the Medina Gazette. This was also indicated by the remedies used, linseed and kerosene oil. Where this was used in season a cure was effected. This Texas fever was the only disease that I know of that has been epidemic among cattle here. Sporadic cases of various diseases have appeared now and then, but we are pretty well versed in such diseases and know what remedies to use.

Mr. FRED. P. NEWKIRK, Oxford, Chenango County, New York, says:

In reply to your inquiries, I would say that abortion in cows and black-leg in calves are the principal diseases in this vicinity. Not keeping cows, I can give you little or no information in regard to abortion. Keeping about one hundred calves, I think I am well posted in regard to black-leg. A calf attacked with it will be stiff in the limbs; its eyes will sink in the head and it will lose its appetite. The duration of the attack is from two to twenty-four hours. Sometimes they will live two days. I have never heard of one recovering. After the discovery of an attack of black-leg the animal is as good as dead. If bled no blood will flow. The disease usually settles in the legs, hips, or shoulders. The exact locality can be ascertained by rapping on the animal with the ends of the fingers, and when the affected parts are reached the sound will be like that produced by rapping on blubber, and, in fact, when cut open the part affected will be found black and blubbery. I have cut a slit four inches long in a shoulder without a sign of distress, the part affected being entirely dead.

In those that I have dissected the internal organs were found perfect, but the heart and veins were full of coagulated blood as black as tar. The fact that little or no blood can be drawn from the large vein in the neck of the calf attacked is to me conclusive evidence that the disease is one of the blood.

I think bleeding is a preventive; at any rate, I lost three with the disease in as many consecutive days, and I immediately bled the balance in the neck, taking about a quart of blood from each, and have not lost a calf since. A neighbor of mine, keeping twenty-three calves, lost six during the fall months with the disease. He bled the balance and lost no more. The calves are usually attacked after being turned on " after-feed," and should be bled before, and again in October.

Mr. T. S. GILLILAND, Van Wert, Juniata County, Pennsylvania, says:

Hogs are suffering here from what is known as hog-cholera. They are mostly taken with a cough, and some die in a few hours, while others linger for three or four days, sometimes for a week or two. When opened the lungs seem to be very much affected and have a very offensive smell. Some persons claim that they can detect this smell while the hogs are yet alive. They seldom recover. Some apparently get well and gain in flesh, and then die. One man brought me a portion of the fatty part of a hog that he said had had the cholera and recovered, and had fattened as well as the rest of his hogs ; but the meat was a bright yellow. I thought likely this discoloration was caused by an obstruction of the gall-duct, so that the gall had disseminated itself like in jaundice in persons.

The cholera seems to be epidemic in its nature, taking off nearly all the hogs in a neighborhood, while other neighborhoods may entirely escape. Changing hogs from one place to another seems to be beneficial. One man had his hogs in pens, and after losing between forty and fifty turned them in a woods lot, after which he lost no more. He thought he had discovered the cause. Another farmer had his in a large field, and after losing about thirty he put the remainder in pens, and they did well. He thought he had found the cause and a remedy. Some of our physicians claim that the disease is lung-fever, while others think the affection of the lungs is not the first cause.

Chickens also have the cholera. They seem to have a diarrhea. Some will linger for three or four days, while others, which seem to be in apparent good health in the evening will be found dead under their roost in the morning. Cleanliness of coops and roosting-places seems to have a good effect, but is not a sure preventive. Some persons claim that the disease is caused by chickens becoming lousy and eating the lice. It is claimed that common black pepper and capsicum administered in sour milk is both a preventive and cure.

Dr. C. H. E. SHUTTEE, West Plains, Howell County, Missouri, says:

A few hogs have died in this county of a disease called cholera. I do not think the disease was cholera; it seemed to be more of an affection and inflammation of the lungs than anything else. I do not know of any remedies that were used, as the disease prevailed to so limited an extent as to attract but little attention. There are no diseases among other classes of farm-animals in this county.

Mr. JOHN HORNBACK, Carthage, Jasper County, Missouri, says:

We have had no prevalent diseases among horses since the epizootic, four years ago, except the common horse or colt distemper, which seldom is treated with medicine or proves fatal.

There is and has been a fatal disease prevailing among the cattle of the county. It is known as Texas or Spanish fever, and is very fatal. During the past summer, wherever Texas or Southern cattle were herded nearly all the home or native stock of cattle died off. In some neighborhoods and settlements there are scarcely any cattle left. When first taken they appear to droop around for a day or two, looking very gaunt and hollow. They also have a hot fever, with little or no appetite. About the third day they appear to fail very rapidly, and in many cases do not live beyond the fourth day, and rarely if ever longer than the sixth. If examined after death the stomach or manifold, and the food contained therein, will be found as dry as dry light wheat bread, and the folds of the stomach will be about as tender as wet brown paper. There are many reported remedies for the disease. I have tried a great many of them myself, but have never succeeded in curing a single animal. I think the only preventive or remedy is to keep the Texas cattle away from our native stock. Our cattle never have the disease unless they run with or are grazed on the same pasture with Texas cattle.

For the past two years we have suffered to some extent with diseases among hogs. The disease is called cholera by many persons, but instead of but one I think there are many diseases. During last summer and fall a great many pigs and shoats died. Some would die in a very few days after being taken sick, while others would linger along and live for nearly a month. It is my opinion that most of the pigs and shoats that were lost died from the effects of worms. Soon after death a great many small worms would crawl out of the nose and mouth, and when cut open and examined the stomach and lungs would be found infested with large numbers of small, white, wiry-looking worms. I am of the opinion that some of the larger hogs died of the genuine hog-cholera, but I have heard of no certain remedy for it.

Nearly every summer a disease prevails among fowls in this county. I think the disease is what is generally known as chicken-cholera. We have no certain remedy for it.

Mr. HENRY WAYMIRE, Little York, Montgomery County, Ohio, says:

We have had no epidemic among horses since the prevalence of the epizootic some years ago. I hear of no diseases among cattle and sheep, and presume both classes are unusually healthy. There has not been so much cholera among hogs this year as formerly. There have been but very few cases in my neighborhood. Wood-ashes, fed in slops, are used as a preventive. Turnips are said to be not only a preventive, but also a cure. We lose a number of fowls every year from a disease known as cholera. Coal-oil mixed in their feed has proved quite a good remedy.

Mr. T. J. CONOVER, Monroe, Butler County, Ohio, says:

I have had considerable experience with diseases incident to hogs, and ever since cholera has been in our land I have been endeavoring to find a cure for it. I have tried many preventives and cures recommended by journals, &c., but found none of them to be certain remedies. For the last two or three years I have proven by my own experiments that the process of changing from field to orchard, meadow, woods-pasture, or roadside, or any new place will be attended with favorable results. At the first appearance of the disease I begin this changing process. I watch the hogs and whenever they come back and lie around the place of entrance I give them a new place, and continue to do so through the day as often as I think necessary. I feed them no grain, but give them all the slops from the house. My theory is that the well ones will survey the new place and the diseased ones will follow them around. This exercise induces a circulation and warms up the system. What grass and herbs they get will be found good for them. Now for the proof: In July, 1876, I had some ninety pigs, and out of this number saved seventy-five, which remained in good condition until proper age for market. One of my neighbors, who had 120 head, saved but four. Two others, who had over 100 each, saved but eight, and so on through a long list. My neighbors were trying different experiments with various kinds of medicines, while I was practicing the changing process. Thousands of dollars have been fruitlessly expended in the use of medicines, from which no benefit whatever was derived.

Last May my pigs were affected with cankered sore mouths and noses. Their mouths were so sore they could not nurse, and they were in an almost starving condition. I took them from their mothers, put them in a clean, dry pen, with good bedding, cleaned their sores, and applied grease to keep the scabs soft. I then fed them on fresh milk with a little water in it, and they soon recovered. Pigs, if taken in time and treated in this manner, will generally recover. As to the cause of this disease, I have no knowledge.

Mr. JEREMIAH CHADWICK, Smethport, McKean County, Pennsylvania, says:

There are no diseses prevailing at present among farm-animals or fowls in this county. A few, and but very few, horses have died in the oil localities of epizootic. I lost four head of young cattle with black-leg, and have heard of two or three other cases. The disease and its causes and remedies are so well known that I will not attempt to throw any new light on the subject.

Mr. JOHN N. GEARHEART, Troy, Miami County, Ohio, says:

A disease is prevailing among hogs in this county which is commonly called cholera, but it appears to be more like a lung disease. They have widely different symptoms. Some cough, some have high fever, some are lame, some bleed at the nose, some are very thirsty, and all seem to lose their appetites. Soda, soft-soap, wood-ashes, and cracklings have invariably proved good remedies for my hogs. I lost two or three hundred dollars' worth of swine before resorting to these remedies, but since using them have lost none. A farmer near me, who had lost quite a number of hogs, commenced to give one pound of soda in slops to 50 head of hogs twice a week, and has since lost none.

Mr. John Gordon, Lynnville, Morgan County, Illinois, says:

With the exception of a disease among fowls, and the so-called cholera among hogs, our stock is and has been reasonably healthy. But our farmers are annually great losers by the ravages of the disease called cholera among hogs. The loss in Morgan County is estimated at twenty-five thousand head annually, worth, on an average, $10 per head, aggregating $250,000 per annum—a loss largely greater than our farmers can well bear. There is every reason why Congress should make an appropriation to enable your department to investigate the cause of the disease and the remedies necessary to cure it. I am satisfied that if a like disease prevailed among the food-producing animals of Europe, millions of dollars would be expended in efforts to discover the cause and remedies to prevent and cure the disease. It is too expensive and extensive for individual enterprise. While we have had the disease on our farm five different years in the last fifteen years, I am as ignorant as to its cause and the necessary remedies as when it first came. It seems to come at all seasons of the year, and the hogs are generally operated on differently. Many remedies have been tried on our farm, but as yet without beneficial results. The causes are so obscure, and the treatment is, as far as I know, so unsatisfactory that it is difficult to give anything like a clear statement on the subject. I am of the opinion that a commission composed of scientific men, employed indefinitely, would in time discover the cause and with it the necessary preventives and remedies. I hope that you will at an early day call the attention of Congress to the importance of the subject, and that it will make an appropriation sufficiently large to investigate the whole matter thoroughly.

Mr. J. E. Karr, Big Flats, Chemung County, New York, says:

About the 1st of October last I bought a lot of cattle in the Buffalo cattle market, said to have been raised and fed in the State of Wisconsin, and on the 10th of the same month nine of those cattle were sick and one had died. I sent for Professor Law, veterinary surgeon of Cornell University, and after examination he said they had the Texas fever. I commenced using his prescription to prevent the disease from spreading and to save the sick ones. Out of nine attacked I lost five head. On the 19th of November I bought another car-load of cattle, said to have been raised in the States of Ohio and Michigan. About two days ago the same disease made its appearance among them, and how many of them I shall lose time will tell. Now, what I wish to say unto you is this, as you are at the head of the Department of Agriculture you might lay such facts as these before Congress, and ask it to enact some law to prevent the spread of the disease by prohibiting the transportation of Texas cattle to the East. These cattle are brought to the markets of Chicago, Detroit, Buffalo, and indeed all the great cattle markets of the country, where they are fed and watered, and the next day cattle from other States, or native cattle, as they are called, are brought in, and, eating the hay the Texans left, get the disease and spread it all over the country. No one is responsible. Farmers who go to the markets to buy cattle are not to blame, as they do not know what yards sick Texans have been in, and dealers do not care as long as they can sell and get their commissions. I do hope you will try to get Congress to do something toward prohibiting the shipping of diseased Texas cattle through the country.

Mr. A. Coffman, Reynolds, Rock Island County, Illinois, says:

At present the only prevalent disease among farm-animals here is cholera among hogs, and of this there are so many different forms, that it is difficult to give a diagnosis of it. It not only occurs in widely-different forms, but also under circumstances and conditions as varied and as widely different. Hence no theory has yet been advanced here but that some well-known facts occur which knock the theory "higher than a kite."

The form of the disease which prevails here to the greatest extent, and which causes the greatest loss to hog-raisers, is what is termed pig or shoat cholera. I should say that it resembles a low form of typhoid pneumonia, generally attended with a violent cough, sometimes with vomiting and purging, frequently with sore head and eyes—the eyes sometimes bursting entirely out of the sockets. They sometimes live for weeks, all the time wasting away, and occasionally die within a few hours. This form seldom attacks hogs a year or more old. The more violent forms vary so much, that I will not attempt a description. As to the remedies, they are as varied as the notions of the owners can make them. Everything that is heard of or can be thought of as likely to be of benefit is tried, but as often fails. My own experience is (and I have had considerable of it) that medicine is of little use. I had it among my shoats last winter, had previously used nothing to prevent it except a little concentrated lye occasionally (if that be a preventive), and used nothing while it lasted in the way of medicine. I changed their rests every other day, and had them driven considerably every day. Under this treatment I lost but few, and escaped better than my neighbors. Still I do not advance this as a sure means of cure. I have more faith in it, however, than in all the drugs of the apothecary combined. Others have tried the same treatment to some advantage. The disease is very destructive here again this winter. I sincerely hope, with the combined efforts of yourself and the stock-raisers of the country, that some preventive may be found for this scourge.

MITCHELL BROTHERS, Hannibal Centre, Oswego County, New York, say:

There has been quite a heavy loss here incurred from a disease among hogs. We have no name for the disease, as there seems to be no definite knowledge concerning it. Some people call it "black teeth." The first symptom noticeable is lameness in their hind feet. This continues until they lose the use of their hind legs entirely, after which they soon die. They have but little or no appetite after they are taken sick. There have been a great many hogs lost by the disease in this neighborhood during the past eighteen months. We lost five head ourselves last season. We sincerely hope some remedy may soon be found.

Mr. T. P. HAMILTON, Hartford, Fulton County, Arkansas, says:

During the past season we have suffered the greatest loss ever known among hogs in this county. In March last the disease appeared among the pigs and shoats in rather a mild form of thumps. The losses were not very great. In the month of August the cholera made its appearance, and proved very fatal. The greatest fatality was among young hogs. The first symptom of the disease was extreme sluggishness. This was generally soon followed by rapid breathing, sometimes by purging, and at others by blindness. Sometimes they survive for days and at others die quite suddenly. It is not unusual for the flies to blow them before death. The loss has been fully 50 per cent. of those attacked, and of this number at least 10 per cent. have been large hogs. We have no remedy for the disease.

Mr. SAMUEL WIEDMIRE, Grampian Hills, Clearfield County, Pennsylvania, says:

I have lately been informed of the prevalence of disease among fowls in some localities of this county, but do not know the nature of it, nor any of the remedies or preventives made use of. When disease makes its appearance among swine, the principal remedies are charcoal and sulphur. As far as I have been informed, most diseases among this class of farm-stock yield to this kind of treatment. Some years ago I lost a few hogs myself, but I believe the trouble was caused by keeping them too long on a plank floor during the winter season and feeding them principally on hard corn. I find they always do better where they have pens so constructed that they can have a good-sized yard in which to exercise.

Mr. C. GINGRICH, Reisterstown, Baltimore County, Maryland, says:

A disease has been prevailing among cattle in the vicinity of Baltimore for the past twelve or fourteen years, and in many cases has proved fatal. As most of the cattle in this district are milch-cows, the disease prevails most extensively among them. It is commonly called lung-fever, but as it is identical with pleuro-pneumonia, it should perhaps more properly be called that. It has thus far baffled all medical skill. It seems more malignant where a large number of cows are confined in filthy stables. I know of several dairymen who were compelled to suspend their business on account of heavy losses by the disease. Renovating the stalls, whitewashing, using carbolic acid, carbonate of lime, and smoking the stables with tar, &c., have had the effect to check the disease for a time, but it is liable to break out again. The symptoms are a cessation of the milk secretion, loss of appetite, and stupor, accompanied with quick pulse and high fever, and secretions from the nose and mouth. Some animals die within a few days, while others linger for some time. Fresh cows are more liable to attack than dry ones. Nearly every case proves fatal. The disease is undoubtedly typhoidal in its character. Some years ago a bill was introduced in the legislature providing for an investigation of the disease, but it failed to pass.

There is no class of animals among which such heavy losses occur as among swine. There is certainly something wrong in the rearing and management of hogs, as the losses sustained amount to millions of dollars annually. I am of the opinion that the cruel system as now and for many years practiced has a great deal to do in inducing disease among this class of animals. The hog is an animal that cannot endure such hardships as horses and cattle. In the Western and Southern States swine diseases prevail to an alarming extent. In these States a most cruel and injudicious system is practiced in the rearing of the animals. Raised without shelter either from the burning sun of summer or the cold storms of winter, it should not be wondered at that they contract disease and die by the hundreds and the thousands. Young shoats should not be fed entirely on corn, as this feed produces an abnormal growth which results in a weakened vitality, and generally ends in cholera or some other disease incident to these animals. In the Eastern States and other localities where the hog is raised under a better and more careful system, cholera and other diseases are not known. I believe if a better system were adopted that cholera and many other diseases to which the hog is now liable would be almost entirely banished. We have sustained heavy losses in this county, and must continue to do so until a change is made for the better in the treatment of swine. So long as hogs are confined in dirty, filthy, muddy pens, and fed on nothing but dry, hard corn, we must expect them to sicken and die. The cholera, so called, is also typhoidal in character, and this opens up a wide field for investigation.

A disease prevails among the poultry of our county which has been very destructive. Some farmers have lost their entire flocks. No remedy has been discovered. Preventives are used with some success. The poultry-house should be large and well ventilated and whitewashed frequently. The droppings should be removed every few days, and near the door, exposed to the rain, should be placed a good quantity of lime. They are fond of this and will eat it every day.

Mr. JACOB GRUNDY, Lewisburg, Union County, Pennsylvania, says:

Hog-cholera prevailed to a limited extent here last year, but I have heard of none the past season. Fowls are affected almost every year with various diseases, such as roup and cholera, but I think the latter should more properly be called dysentery. The losses are never very heavy. I have found wood-ashes and charcoal a good preventive for cholera among hogs. There is no prevailing disease among either horses, cattle, or sheep.

Mr. B. LE SUEUR, Knoxville, Crawford County, Georgia, says:

We estimate that nearly or quite one-half our pigs die. In portions of this county some farmers have lost one-half, some three-fourths, and a few their entire stock of hogs, and many of them were in good condition for slaughtering. The truth is, many of our people think it too small a business to doctor a hog; and the remark is often heard: "There is no use in doctoring." They seem disheartened from the start. Others change the remedy so often that the medicine kills the animals. As in the West, every disease that is fatal to swine exists here. Many hogs have died without either purging or constipation, and yet the disease was called cholera. One farmer tells me that the livers and kidneys of his hogs were found almost rotten; that the skin was covered with spots as red as blood, and yet they died of cholera. Any information that your department can furnish to stay these fatal diseases among hogs will be highly appreciated by the people here.

Mr. DAVID BRUMBAUGH, Hagerstown, Washington County, Maryland, says:

The greatest fatality that prevails among the farm-animals of this county is among hogs. The disease is what is generally termed cholera. No remedy has as yet been found. The animals affected continue to droop from one to two weeks before death ensues. If the department should succeed in finding a remedy for this wide-spread and fatal disease the whole country will be greatly benefited. Congress could not do a wiser thing than to make a liberal appropriation for its investigation. I had no idea of the extent of the losses in this county until I commenced inquiries in order to answer your letter intelligently. The disease is often confined to the "pen-hogs," during the fattening season.

Mr. C. C. THOMAS, Point Pleasant, New Madrid County, Missouri, says:

So far as my observation goes, the most prevailing disease among hogs is just the opposite to what I understand cholera to be. Their bowels are badly bound up, and in what few I have opened, I found the excrement in hard, dry lumps, and the entrails badly inflamed, with bloody water both in and on the outside of them. Nearly every one so affected that I turned on green clover got well, but when they are attacked in this way while on green clover about one-fourth of them will die. I have had a few that had sores on their faces, but a few applications of carbolic acid, with a little in their food, generally cured them; at least they got well.

Chickens, turkeys, and Guinea-fowls are affected and die in the same way. They mope about and eat but very little; their bowels are very loose, and the discharges often watery and very offensive. They live from seven to eight days after the attack sets in. So far I have been able to find no remedy. I think at least nine-tenths of those attacked die.

Mr. SAMUEL LEA, Leasburg, Crawford County, Missouri, says:

In October, 1876, my swine commenced to get sick, and twelve of them died. They were not all affected alike. All of them, however, commenced by appearing dull and sluggish, refusing food, and moping around. Some had a cough and difficulty in breathing, accompanied by a very feverish condition generally. Still others had a diarrhea, and their evacuations were very black and offensive to the smell. But one of those attacked got well. I found her hungry, and gave her about thirty grains of calomel in wet corn-meal, which she ate. I have no idea as to how the sickness originates. My hogs were thoroughbred Berkshires, and did not come in contact with other animals, and their feed and water were good. No water on my place flows on to it from other lands. I fed wheat middlings, and they had the run of a clover pasture, orchard, and unbroken woodland. It is now over twelve months since I lost a pig. I have given the same feed and treatment as last year, and have several of the same animals I had then.

Mr. WILLIAM A. BULL, Frohna, Perry County, Missouri, says:

We have a disease prevailing among our hogs which I will try to describe. In July last it made its appearance about six miles northeast of me; it is within one mile of me now. The hogs are still dying with the malady. The duration of attack is about two weeks for grown hogs, but for small pigs about two days. The average fatality is ninth-tenths. When first attacked the hogs get lame, apparently as in cases of rheumatism, have a dry, hacking cough, are costive, and have excessive thirst. Some have small sores on their legs and ears; some are partially and some totally blind. In the last stages of the disease they are purged severely. Among the remedies used I will mention calomel, turpentine, sulphur, copperas, tar, poke-root, and many others, but all without success. I have not had an opportunity of examining any of the hogs after death. They have been differently situated when attacked—in stubble, clover, and woods pastures, and some in pens.

Mr. MARTIN J. SACKETT, Houseville, Lewis County, New York, says:

This is a dairy county, and there is no prevalent disease except among cows. We have lost heavily from garget in the udder of cows. Poke-root sometimes helps it, but not often. Lumps very frequently come also in the teats of cows—a sort of stoppage—which has been a source of great loss to us. We know of no remedy.

We have suffered also from abortion among cows to the extent, in some instances, of one-half the herd. It has not been so prevalent, however, the past three years as it was previous to that time.

Dr. JOHN M. McGEHEE, Milton, Santa Rosa County, Florida, says:

I know of no diseases affecting horses in this section which do not prevail in other localities generally, and I will only mention some remedies which are new, so far as I know, and some circumstances connected with those diseases not generally noticed. The greatest fatality seems to result from colic, and a new remedy, which has been used in this section of the South when all other remedies have failed, has been to perforate the walls of the abdominal cavity at a point just half way between the prominence made by the hip-bone and the ribs. This remedy, however, has been recommended by some modern works on farriery. While on this subject I will here mention that before the war, on some large cotton plantations, I noticed that nearly all of the mules and horses which died of colic died on Monday, a very few on Tuesday, and a still less number on other days of the week. These facts I think point clearly to a cause and a remedy. The animals being worked all the week in warm weather, their exercise brings about a certain degree of digestion and appetite. For lack of exercise on Sunday their digestion is weakened, and in most instances colic is the result on Monday. It is plain that the remedy for this is to reduce the feed on Saturday night and Sunday.

The next most fatal and common disease affecting horses in this section is that known here as "blind-staggers." It is first discovered by some foolish or unaccountable act of the animal, and as it advances the intelligence and control of the muscular functions become more clearly affected, until the animal seems to be frenzied. Death generally ensues from within twenty-four to sixty hours. Examination of the brain shows extensive inflammation and serous effusion. The only remedies in this disease in popular use, which are relied on, seem to be herculean and to some extent empirical. The first remedy which I will mention is to bleed profusely on the discovery of the first symptoms of the disease, and then give a dose of spirits of camphor, spirits of turpentine, and tincture of asafetida, and whisky. All of this seems very contradictory, but it is confidently relied on by many who have witnessed its effects. I saw it tried once in an advanced stage of the disease—too late for any remedy to do any good—but in two or three minutes perspiration poured out from every pore. I think if there is any good in this dose it is owing to the almost caustic and destructive effect of turpentine on the flesh of the brute creation. I know that bleeding alone will not arrest the disease. The other remedy is to take two switches, sharpen and introduce them through the nostrils in the region of the brain, then give them a thrust and pull them out, when the blood is said to flow freely, and if used in time the horse recovers. I heard one man say that he once tried the remedy, and the horse fell as dead as if his brains had been shot through with a ball. The preventives are simple and sure. They are simply sound food. I know that damaged grain will produce it, even if the damage is so slight as not to be readily discovered, as new-ground corn often is, or shipped corn, slightly heated from incipient fermentation, or late corn affected with smut. My experience on these points is full. In 1836 I lived in Montgomery County, Alabama. The corn and cotton crop that year was a failure. Most persons got their corn from New Orleans, which had been shipped down the Mississippi River in flat-boats. The corn generally looked well, and when planted came up, but much of it was damaged and great numbers of horses died of "staggers." I was a boy then, and I heard it attributed to the shipped corn ; and I have never known a case of "staggers" which I could not trace to some of the above-mentioned causes. A few years ago I sold some corn to two log-men for the use of their oxen. They had one horse each, and both were valuable animals. I knew the corn had been heated, and in the most urgent manner cautioned them against giving it to their horses. But they fed their animals with it, and both horses died of "staggers." Some pastures at certain times are said to produce the disease. In such cases it would seem that it is caused by a web on the grass.

The following remarks on the treatment of horses will perhaps be new to many : About twenty-three years ago I had a horse very badly foundered. I tried various remedies of empirics, but my horse grew worse. After witnessing his sufferings for several days I resolved to know what the founder was (having a contempt for such works on farriery as I had then seen). I prepared to cut into and lay off the whole of the thin covering of the bottom of the foot. Setting my knife obliquely to avoid puncturing the capillaries as much as possible, I introduced it at a point between the frog of the foot and the toe. As soon as I punctured the thin horny covering a serous-looking fluid was emitted. I extended the incision far enough to examine the integument underneath this covering. I discovered the mucous covering of the capillaries entirely separated from the horny portion of the foot. The vascular portion of the foot was highly inflamed and as sensitive as an exposed nerve. I then cut into the other foot and let out this serum. It seemed to me to be analogous to an ordinary blister on the skin, where the cuticle is lifted up and leaves the mucous coat or serum intervening. This serum showed no disposition to harden on exposure to the air and stop up the orifice, but continued to be as limpid as oil of turpentine. He was a long time recovering, but when he did recover the cure was radical. The foot was not the least affected. After he passed out of my possession I continued to inquire after his welfare. He never foundered again or complained of his foot. Had this lymph remained in the foot it would have formed a fungus substance, which would eventually have produced what is known as chronic founder. This operation should never be performed until several days after the founder is known to have caused this effusion. The incision should not be large, and should be made very oblique in order to cover the integument. After the inflammation had entirely subsided the exposed parts were very tender, and I had thin, solid shoes put on, which covered the entire bottom of his feet, and he traveled without any difficulty.

I have had several horses foundered since, and I never found any difficulty in curing them in twenty-four hours by fastening around their ankles cloth or rags and pouring warm water on the bandages. I have generally carried it so far as to produce blistering of the ankles, which has sometimes been slow in curing up. This remedy should be applied as soon as the founder is discovered, and before the formation of the serous discharge. In no case should the horse be used for several days. I believe that oil of turpentine would answer the desired end if used on the ankle after several hours' use of the warm bath, applying after the hair is wiped dry.

I have observed in horses a very marked tendency to metastasis when diseased. This peculiarity may account for their susceptibility to the action of counter-irritants.

All other agricultural interests sink into littleness when compared with cattle-raising. In the Gulf and South Atlantic coast it is blended with our hygiene and civilization, and yet it is hard to find an example of any interest so much neglected. Perhaps it would be best to first explain the cause of this neglect. In 1836, and a short time after the last Indian insurrection of the Creek Nation, and a short time after the massacre of the stage passengers, and the burning of the stage and United States mails, I traveled through the Creek country from Columbus, Ga., to Montgomery, Ala. Sixteen miles from Columbus were found the bones of the stage-horses and some of the charred wood of the stage-coach. Every white person had moved out of the nation. The public mind was greatly excited, and I was left alone with thirty or forty negroes and eight horses to wend my way through the country. There was no corn to be had until we reached the station of the United States soldiers, where ample food was obtained for our horses. Having passed the Creek territory I reached the lime lands of Montgomery, where I saw some cattle, and they continued to increase in numbers and size the farther I progressed into the prairie lands. These cattle had very much the appearance of the Texas cattle now. In the years 1836 and 1837 the Indians were moved out, and farmers from Georgia and the Carolinas soon occupied the lands thus vacated. The grass in the summer and cane-swamps in winter kept their stock in fine condition. The cows had calves every year, and soon the woods were teeming with cattle. Almost without care or feed the cows produced an abundance of milk and butter the year round. But a change was all the time taking place. The large herds while feeding on the hill-sides were cutting the grass roots with their feet and loosening the soil and sand. The rains would wash this earth and soil down into the edge of the cane-swamps, giving the stock a foot-hold to reach the cane on the edge, which was otherwise inaccessible. And so steady and rapid was this change that eight years after, when I traveled through this section of the country on the same road, those cane-swamps were marked only by sandy branches with some switch-cane on the edges. The cattle-range lasted much longer near the Gulf coast, for the reason that the country is more level and generally less inviting, and is farther from the sources of supply and population. It was under these circumstances that cattle-raisers formed their habits. These surroundings lasted a full generation, and a generation grew up who knew no other resources. Cultivated pasturage and hay-lands are unknown to them, and as I have repeatedly reported before to your department, I do not know of one acre of ground cultivated in West Florida for pasture and hay, though my acquaintance is very general. That you may be enabled to form a correct opinion of the losses of the cattle interest in this section, I will give you the system of stock-raising here. About the 1st of February each year much of the grass range is burned off, and all the young and tender shrubs which grew up the previous summer are killed by the fire. The wire-grass first starts to grow and puts forth large bunches of young, tender, and quite nutritious growth. This burning is done only in small spots of a few miles square. The cattle soon find it out and gather on "the burn." This burning is done to draw the cattle from the low lands, for at this season they are very poor and weak, and hardly able to get out of the smallest bog. The weather is generally such that, if there is much rain, the cattle catch cold if they lie down at night. Many become stiff and lame, and some are never again able to rise to their feet. If they are lifted up they are so far exhausted that they rarely recover. The remainder of the loss is in boggy branches, where the cattle reach after a little green switch-cane. These losses frequently amount to 80 per cent. of the breeding cows and a much smaller per cent. of the dry cattle ; but as the heifers are rarely ever sold or killed for beef, the stock is thus replenished. If from accident or otherwise fires get started and the woods are burned while the weather is too cold for the grass to grow fast, many cows gather on "the burn" and perish while nibbling at the short herbage. This short grass is very weakening to them, as it inclines them to scours.

The general burning of the woods is about the last of February. The weather at this time is usually warm and the grass shoots up rapidly. The cattle recover very rapidly, though for a few days they suffer much from hunger, as the whole country is a charred waste.

About the 1st of April the cattle-owners appoint a time and place of meeting to make a "drive." All the cattle at pens are collected and the owners separate them. After they are separated, the large stock-owners drive fifty or one hundred five or six miles away from any other large body of stock and give them into the care of stock-minders, who have pens built for the separation of the cows and calves, and whose compensation for this service is the milk from twenty to fifty cows and the manure from twice or three times as many dry cattle. The cows average about one quart of milk per day. Since the partial destruction of the range the cows have calves but once in two years. Very little milk is taken from what are called the calf-cows, the most of it being taken from the yearlings. All the cattle are supposed to be penned every night, and from one to three acres of ground is what is called "trod." On this "trod" land the stock-men plant corn and sweet-potatoes. · Some plant a small patch of sugar-cane. About the last of July or the first of August the calves are all marked and branded, and the whole herd is turned loose to hunt the wild oats on the unburned spots of the early spring. They have free range to gain all the strength they can to take them through the winter. Beef so raised is not good. It has but little flavor, and persons who are able to buy good Texas or western beef will not buy it if they can get the better. Some few steers eight years old or more make very fine beef late in the fall.

A moment's glance at this manner of cattle-raising will convince you of the severe trials infants and children from one to three or four years old must undergo. Four months of the year they have an abundance of milk, a food easily digested and answering most of the demands of the animal economy. They use little else. Their stomachs are fitted to its easy digestion. In an hour it is all taken from them, and the most indigestible of all food for children is substituted. Now, new sweet-potatoes, corn-bread, and pork or bacon is the food for their tender stomachs to digest. The change is too great for their delicate organs of digestion. They feel a restless craving for something, and they eat whatever comes in their way—rags, paper, pine-bark, rotten wood, and finally the clay with which their chimneys are daubed.

Sheep in this section, like cattle, suffer from few diseases except such as are brought on from neglect. The scab is a common disease among them, and so far as I know but few attempts are made to cure it. It is also a very common disease among goats. Sheep likewise suffer from rot. I have recently tested tobacco as a remedy for this disease. Sheep eat it very readily when it is mixed with their food, and soon become fond of it. If properly used, I think it will effect a cure. The range for sheep is much better than it is for cattle, and they generally keep in good order most of the year.

I have had a great deal of observation, but little experience, with the diseases of hogs. During the war a good many hogs died of cholera, for which there was no remedy. Copperas was used as a preventive with some success. Preventives I believe to be the only safe policy.

In conclusion, let me assure you that there is great room for improvement in agriculture in this section, and much can be done by your department if it is afforded the necessary means. It is pleasant to witness your efforts to build up the substantial interests of the nation, and your confidence in the prospective economy is not a vision.

Dr. J. G. Hart, Murray, Calloway County, Kentucky, says:

A disease uniformly fatal to horses has prevailed in this section for two years. It appears to be propagated by actual contact with matter or virus, inasmuch as animals kept separate though near the disease are not liable to take it. Some regard the disease as cold distemper, while others believe it to be glanders. The symptoms are about as follows: At first fever, which is soon followed by a dry cough and a nasal discharge resembling that from ordinary distemper. There is more or less enlargement of all the glandular organs so far as can be observed. Constitutional disease soon sets in, which is denoted by the change in the nasal discharge from a watery to a gleety and offensive flow. The animal loses flesh rapidly; the skin soon becomes thick and eruptive; the lymphatic glands throughout the body become much enlarged, but never soften or suppurate; the submaxillary and sublingual glands are most especially involved, at least in most cases to the extent of suppuration and softening. The duration of the disease is from two to twelve months. It is invariably fatal. Quite a number of remedies have been used, but without success. Veterinary surgeons have been employed with like ill success.

Hog-cholera, with its usual symptoms, has prevailed to a considerable extent in this locality. The only remedies that have been used with any degree of success are hygienic. If the animals are confined in a dry lot when the disease makes its appearance, the mortality will be very small. Drinking cold water appears to be the immediate cause of death in a majority of cases.

A disease called by some hog-measles prevailed here as an epidemic from June until October of this year. The disease is characterized by a high grade of fever for two or three days, which is followed by an eruption about the head, neck, and shoulders, and in some cases of the entire body. This lasts three or four days, when the animal either begins to recover or dies. Sulphur and poke-root have been used with apparent good success. Hogs should be confined to prevent them from drinking too much. The fatality corresponds with the character of the epidemic as to mildness or malignancy.

A disease called chicken-cholera has also prevailed quite extensively. By commencing early I used carbolic acid with good success for two years, but it has signally failed this year. Confinement in coops elevated above the ground, with little or no water to drink, would seem to be the surest remedy.

Mr. H. P. Jordan, Victoria, Victoria County, Texas, says:

Native cattle are free from disease and comparatively healthy, but I think fully one-third and perhaps one-half of all the Durham cattle imported into this section of the State have died during the past two years from what people are pleased to term acclimating fever. The disease appears to be similar to that which afflicts the cattle in Missouri and Kansas, and which is supposed to be imparted to them by the native cattle of this State. I think all cattle brought here have this disease sooner or later. The first symptoms of the disease are fever and constipated bowels. The principal remedy used is castor-oil. The disease is a very serious drawback to the cattle-raisers of this State, who are trying to improve their long-horned Spanish breed with short-horns. If anything can be done to arrest it, great benefit will result to the people of this State. One of my neighbors lost four out of six fine Durham bulls, another lost three out of five, and a few have lost all.

No unusual disease exists among horses and hogs. Chicken-cholera prevails to some extent at times. There are a great many cures recommended, but I think all of them fail when the disease gets a good start.

Mr. JOHN W. GILL, Clay County, Missouri, says:

Hogs die here by the thousands of cholera and are doing so all over the country No certain cure has been found. I have used a great many things, and if anything has done any good at all it has been spirits of turpentine given in slops. Fowls die quite rapidly of cholera. I have used wheat bran and epsom salts as a preventive with good success. Dissolve the salts and wet the bran with it and feed. If they will not eat, drench them with salts—a teaspoonful at a time once a day. Since adopting this treatment I have lost but few fowls.

Mr. H. G. KERNODLE, Kirksville, Adair County, Missouri, says:

A few cows have recently died here of a disease called "mad itch." No remedy known or treatment given.

Over four hundred head of hogs have died of cholera in this vicinity during the past two years. Farmers have used every remedy known, but without success.

Two years ago about all the chickens died. No remedy used or treatment given. This statement relates only to my own immediate neighborhood of about two miles square or less.

Mr. G. W. RAUDABAUGH, Celina, Mercer County, Ohio, says:

The only disease from which serious losses have been sustained is from cholera among hogs and chickens. The disease has prevailed quite extensively among hogs the past season, and on some farms is still prevalent. Two years since it prevailed in a mild form, and about 20 per cent. of those attacked died. This season it was more extensive and fatal, and the losses were about 50 per cent. of those affected.

We thought we had a remedy for the disease, and in many instances it seemed to check it at once; but the past season it failed to bring the expected relief. The prescription for fifty hogs is as follows: Two pounds black antimony, seven pounds copperas, five pounds sulphur, and two pounds saltpeter. Two years ago my neighbor gave this remedy to his hogs after he had lost twenty-five out of a herd of seventy-five, and he lost no more. Notwithstanding the same remedy was given to about one hundred hogs this season, about one-half of them died. Two years since my hogs were attacked by the disease. I gave them no remedy, but removed them about three-fourths of a mile from their old haunts into a woods-pasture, and they all recovered. This season they were attacked in October, and out of fifty head about thirty died.

The disease is always more fatal among pigs than among older hogs. The symptoms are not always the same. In the first stages food is taken very reluctantly and does not seem to be relished. Indisposition to move and general stupor follows; a cough sets in, which I think is caused by a nauseated stomach, and a great disposition is manifested to lie on the belly. In a few hours after death in almost every instance the carcass becomes wonderfully swollen. All things considered, this is one of the most difficult diseases to understand that animals can be afflicted with. My hope is that your investigations may result in the discovery of at least a preventive, if not a permanent cure, for this terrible scourge.

Mr. WILLIAM B. ARNES, Warrensburg, Johnson County, Missouri, says:

With the exception of hogs and fowls, our domestic animals are quite healthy. Fowls are subject to a disease called cholera, of which I will speak hereafter. Within the past two years and a half the farmers of this county have lost heavily by a disease among swine erroneously called cholera. During the time indicated the disease has assumed three different forms. The first, by which I lost most of my herd, constipation was developed. The evacuations were dark and dry. The animal had a feeble, staggering walk, and appeared in great pain. Death ensued within from six to ten hours. When the weather became cooler the symptoms changed. The bowels were loose, there was slight bleeding at the nose, and the urine was strongly colored with blood. *Post-mortem* examinations showed congestion of the lungs, and sometimes worms in the intestines. The last type of the disease is now prevailing among a number of herds in this vicinity and is proving very destructive. Occasionally the hair of the animal nearly all comes off and the skin is a broad, raw surface. I have had two hogs to recover from this form of the disease. I believe the best treatment is to give the hogs plenty of room to range over. I keep salt and ashes in my feed-lot and give them all the pit-coal they will eat, and occasionally a bran-mash wet with poke-root tea. I believe poke-root to be a preventive of disease in hogs. So far as my own experience goes, I have found it a cure for diseases among fowls. By following the above treatment, I have not lost a single hog or fowl by disease this season.

Mr. JAMES W. GRACE, Watterborough, Colleton County, South Carolina, says:

The past year has been an unusual one with hogs. They have been attacked with a disease known as cholera, which usually kills them within about ten days after the first symptoms make their appearance. They refuse to eat and seem to desire to lie down all the time, and apparently suffer very much. I have not known anything like it in the last fifteen years. It made its appearance about the 1st of September.

Mr. AARON DRESSER, Hardinsburg, Breckinridge County, Kentucky, says:

The disease common among hogs, and known as cholera, has prevailed extensively in some parts of this county, and has been very fatal. I have heard of no remedy that can be depended upon.

Mr. C. B. RICHARDSON, Henderson, Rusk County, Texas, says:

Before the war I lived near the Mississippi River, in Carroll Parish, Louisiana. A disease called cholera broke out among the hogs. It was the first epidemic ever seen by the planters in that vicinity. Most of the planters had very large herds of hogs, as there was a good range in the swamps back of the farms. Every form of treatment was used without any marked success. The attacks of the disease were quite sudden. Some would swell up and the flesh would look livid, and they would die in twenty-four hours. Some were constipated and others would have diarrhea. Fat hogs, as well as lean ones, were subject to attack. I had two killed when first taken, and got my family physician to assist me in making a *post-mortem* examination. The bowels were constipated, and the inflammation of the bowels and stomach was very great. I kept the hogs in a dry inclosure, under the gin-house and cotton-shed. I put tar in the troughs, and fed with corn boiled in lye and copperas water, and poke-root decoction to drink, and used various other nostrums in vogue without success. I burned the hogs that died. One neighbor drove his well hogs four miles into the swamp, and made a man camp with them there, with some success, he thought, as they appeared to die at a less rapid rate.

I have lost some large hogs and pigs this summer with this epidemic here. The disease appears to be a violent fever, and kills the animals in a very few days. I put one fine hog in a lot where it had a good, dry shelter. I tried to doctor it with liquids, but could not tempt it to drink anything. I tried to give it a dose of calomel on a piece of beef, but could not induce it to eat anything at all, and finally gave it up to die. It lay three or four days in its bed, and after awhile it got up and ate a few mouthfuls of corn, and finally recovered without any treatment. I fattened it this fall, and on butchering it I found the lungs and intestines adhering strongly to the sides, and the intestines also tied in lumps with thin ligaments. On the intestines was a large ball four inches in diameter, filled tight with thick matter like dough.

Many nostrums published as cures have been tried with such little success that the farmers now let the disease take its course without attempting to do much of anything. When a hog once refuses to eat, little can be done for him.

Mr. WILLIAM DEARMOND, Irish Grove, Atchison County, Missouri, says:

The disease known as hog-cholera (a term applied to almost every malady that kills hogs) has done great damage in this neighborhood and adjoining communities. It is the same in almost every instance; it is only varied by the different conditions of the animal at the time of attack. Here the disease is contagious, and from the time of exposure until its development varies from nine to fourteen days. Symptoms, stupid, and refuse food, high fever, then eruption of the skin, sore eyes, and bowels either constipated or the reverse. In such cases death usually results within from four to seven days. Another form of the disease is that of congestion, producing death within a very few hours. The average fatality is about 80 per cent. There is no effectual remedy known here. As a preventive perhaps complete isolation is the best treatment. I have never known an animal to have the second attack. The disease resembles measles in the human family, and the symptoms are very nearly identical.

Fowls are dying at a rapid rate throughout this and adjoining neighborhoods. The disease seems to be epidemic in form, and kills all on the premises. We know of no remedy, but as a preventive we use white-oak-bark tea, made strong and mixed with corn-meal, and set where the fowls have free access to it.

Mr. C. P. HALLIS, Bloomfield, Stoddard County, Missouri, says:

My hogs are dying at present with a disease that is very fatal. Ten head were attacked, and five died before I commenced treating them. I am now using strong soap-suds, copperas, and saltpeter, and the hogs seem to be improving under the treatment. They are beginning to eat again, and look much better. When first attacked their heads and ears droop, they lose the use of their hind legs, and purge and vomit. They sometimes vomit blood. The disease prevails extensively throughout this neighborhood.

Mr. J. JAMESON, Greene County, Pennsylvania, says:

Chicken-cholera seems to be permanently located here. It has not been so prevalent, however, the past year as previously. Remedies are numerous but not very satisfactory. So far as personal observation goes, I think calomel is used as a remedy with better success than anything else.

Mr. ROBERT W. FRITTS, Lanes's Prairie, Maries County, Missouri, says:

Cattle were quite healthy here until late in the fall, when a few cases of what is generally termed the black-leg occurred. There were some half-dozen cases in my neighborhood, all of which proved fatal. The animal generally lives from twenty-four to forty-eight hours after the attack sets in. Before death stiffness occurs in the hind parts, generally in one hip or leg; the head and ears droop, and dullness and stupor are observed. Fever, and a general quivering of the flesh, especially in the hind parts where the disease seems to be located, also are observable. After death the leg has a black or bruised appearance under the skin. Other parts seem natural except the gall, which appears enlarged. Several remedies were tried, but all failed to give relief.

We have a disease among hogs that has killed about 10 per cent. of them in this neighborhood. Some people term it cholera, some measles, and some lung-disease. It is first discovered by the hog refusing to eat and lying around in a stupid condition. Sometimes they will both purge and vomit, sometimes they will purge and not vomit, or vomit and not purge, and sometimes they will do neither. After death the neck and chest turn spotted, and the insides are often quite pieded. Sometimes they appear nearly rotten; at other times nothing of an unusual character is observed. The duration of the disease is from one to three days, but occasionally a case will linger for a week. After recovery from the first attack, when attacked a second time, a cough sets in, and they usually die in a short time. Those that recover are hard to make thrive or look well again, so it is generally decided here that but little is gained by a cure. Several remedies have been tried—in fact nearly everything that could be thought of—but nothing has proved very successful. I had several hogs attacked, but lost none. I used turpentine as a remedy. I was compelled to drench some of them, but generally I was able to administer it in slops or over their feed. I told my neighbors of it, some of whom tried it and were successful, while others pronounced it a failure. I believe if used properly and in time it is not only a preventive but also a cure. From a tea to a table spoonful twice a day for two or three days is the way I administer it. The disease seems to be contagious, as it is generally from seven to nine days after it makes its appearance among a gang of hogs before others take it, and then dozens may be attacked within a period of twenty-four hours. Dr. Grace, who lost a hundred head by the disease, tried drugs, but finally gave the matter up and considered the malady incurable.

Mr. H. H. CUNNINGHAM, Steubenville, Jefferson County, Ohio, says:

In times past we have had foot-rot, so called, and paper-skin among sheep, and cholera among fowls. Foot-rot to my knowledge has never originated here, but has been introduced by careless handling of sheep brought from other places where it seems always to exist. The localities in which it develops itself without inoculation are in low marshes or moist grounds where the feet are always wet or damp. It is unquestionably a disease caused by wet feet, and a cure without removal from the locality that caused it is an impossibility. The proper preventive would be the drainage of all moist soils, and keep the animal from coming in contact with those already diseased. For a cure the removal to dry soil is indispensable, then the paring of the feet and the application of strong caustics, such as blue vitriol, nitric acid, or butter of antimony. This, with close, careful attention for a few months, will usually effect a cure.

As regards "paper-skin," no cure has as yet been discovered (at least I have no knowledge of any). From my own observation I think it could be easily prevented. It is my opinion that the disease is occasioned by deficient nutrition, as it has always occurred in cold, wet seasons, when pastures are constantly wet and either have some of the elements of nutrition washed out of the grasses, or it may be the lack of heat and sunshine fails to develop those qualities. This, in connection with the unfavorable effects of the weather upon the constitution of the animal, is abundant cause for the low and feeble condition that always precedes this disease, or rather this is the disease itself. A supply of grain in such seasons, sufficient to keep up the normal condition of the animal, would, in my judgment, be a sufficient preventive.

In regard to "chicken-cholera," I would say for this locality that any disease that is fatal to the fowls is so called. I do not know what cholera really is as applied to fowls, and know no remedy. But I do know that the avoidance of close breeding and good care and cleanliness, with healthy food and enough of it, is a sure preventive.

Mr. J. TOWELL, Rankin County, Mississippi, says:

A disease called charbon killed half the horses and mules and many cattle in Rankin County, Mississippi, and vicinity, in 1867. The same disease is reported to have prevailed fatally for the past two years in some parts of Louisiana. This disease partakes somewhat of the symptoms of erysipelas in the human family, being characterized by local inflammation, pain and swelling in some portion of the animal's body, most frequently in the neck, breast, flank, or sides, and is very readily communicated from diseased animals to healthy ones by house-flies, which carry the virus from one to another. But my purpose is not to give a treatise on the disease, but simply to point to a remedy that proved speedily efficacious in nearly every case in which it was employed. Fish-brine is the remedy, and it was used as a local wash to the inflamed parts. Much friction was used and the surface kept wet with the brine until the animal was cured. It is necessary to keep the animal in the shade (stabled) and protected from horse-flies while under treatment. Epsom or Glauber salts were employed internally, given in sassafras-tea when the case was obstinate. Three-fourths of the cases treated yielded readily to the fish-brine wash alone.

Mrs. MARY E. DONLEY, Knoxville, Marion County, Iowa, says:

Hog-cholera has been raging all over our county for several years, and so fatal has it proved that it is regarded as incurable. Many remedies have been proposed and tried with no good effect. The symptoms are, the animal is seized with a hacking cough similar to that of bronchitis, refuses to eat, and turns of a purplish color. I have seen some on our farm where the ears would become badly swollen, and blood would ooze out of them before death. Diarrhea generally ensues. We have thought that in several instances a change of locality abated the disease. Death generally occurs in two or three days; however, in many instances, they are dead before you know anything ails them. I suppose ninety-nine out of every hundred die, or ought to, as they never do any good afterward.

Diseases among cattle are not much dreaded, that known as black-leg being perhaps an exception. I do not know any symptoms; generally find the animal in a helpless condition. If taken in time the disease can be cured by an application of turpentine on the back, over the hips, and on the swollen parts. Must be bathed in with a very hot iron. In fatal cases the animal lives about three days.

Sheep here have become so badly diseased with scab and foot-rot as to make sheep-raising very unprofitable. We find that thorough dipping in tobacco-tea is a certain cure for scab. I have also heard that if blue vitrol is mixed with water and the sheep compelled to walk through it once a day for a few days it will likewise effect a cure.

The raising of poultry is not considered near so profitable as in former years, because of the ravages of cholera. The fowl mopes around or remains on the roost until it dies, which is a very short time. After death the liver is found swollen to about twice its natural size. The heart is also found enlarged. I am sure I have checked the disease several times by using the following recipe: One tablespoonful of finely-ground black pepper, same quantity of alum, and one teaspoonful of soda, mixed in one gallon of sour milk, placed where the fowls can drink as often as they choose.

Mr. JAMES T. DONALDSON, Bowling Green, Warren County, Kentucky, says:

Hog-cholera is the only devastating disease our farm-animals are afflicted with.

Dr. R. BUCKHAM, Phelps City, Atchison County, Missouri, says:

Our loss in hogs by what is improperly called cholera has been very great. In the winter of 1865–'66 the losses amounted to at least $100,000. The symptoms are generally stupor, indisposition to move, stiffness of the joints, eyes weak and watery, sometimes red; constipation of the bowels; discharges black and hard at first. In some cases diarrhœa sets in with bloody discharges and vomiting; high fever and great thirst; occasional bleeding from the nose; in some cases they have cough, in others none; the skin turns a dark purple on the sides, abdomen, and throat. The duration of the attack varies from one day to a month. I think about 80 per cent. of those attacked die. Those that recover peel off like a child with scarlet fever. *Post mortem* examinations reveal the following diversity of phenomena: Congestion of the kidneys; blood in the ureters and bladder. In other cases these organs appear healthy, and the bowels contain a green, degenerate bile of an acid character; in other cases the liver is black in a state of decomposition with empty gall bladder; again the spleen is congested and distended to three times its normal condition. In some cases the lungs show inflammation, with dark spots interspersed through them; and again, in some cases, the stomach contained a green acid fluid, the action of which had destroyed the mucous coating of the walls of the stomach, rendering it not thicker than brown paper. It will be seen from this that different organs are affected in different hogs afflicted with the same disease. I have paid particular attention to the progress of the disease, and I am satisfied it is contagious. In all cases where healthy hogs have come in contact with diseased ones they have been infected. I know of no remedy. We have tried everything recommended, but without success. The only safety is in preventives, and the surest preventive is to keep sick hogs away from the well ones.

A few cattle have been lost, in pens, by a disease which seems not to be understood here. I am told by those who have made examinations that after death dark congested blood is found about the joints. There is no cure that I know of.

Cholera exists among fowls, and it is quite fatal. About 50 per cent. of those attacked die. Smart-weed, cut fine and mixed with dough or given in strong tea, is a good remedy.

Mr. J. E. GRAY, Brenham, Washington County, Texas, says:

A fatal disease commonly called cholera exists among fowls here. The symptoms are first a drooping appearance and disposition to remain on the roost until late in the morning; indifference about food; the wings droop or fall; great thirst, as they drink frequently. Sometimes they show signs of gapes. These symptoms continue from two to three days, when death ensues. When the disease strikes a flock it carries off from 50 to 80 per cent. This season one flock of seventy had but seven left. My wife has used the following remedies with apparent success, but more as a preventive than a cure, viz., red-pepper, sulphur, alum, copperas, turpentine, or rosin, with lime-water to drink. I dissected one that died suddenly, and found the liver in almost a state of decomposition. This leads me to the belief that the liver is greatly implicated in this disease.

Mr. GEORGE A. HYDE, Keating, Pennsylvania, says:

There is no disease among farm-animals in my neighborhood except garget among cows. The remedy is soft-soap and milk of equal parts, one quart every other day until there are signs of improvement. Others give four ounces per day of saltpeter, mixed with pale molasses. These remedies, if properly used, generally effect a cure. Saltpeter and sulphur is a preventive, or in fact anything that will cleanse the blood.

Mr. J. D. SMITH, Greig, Lewis County, New York, says:

A disease made its appearance among cattle in this county in July, 1877, where it still exists. It attacks old and young. The first symptoms are manifested by stiffness and great pain, as the animal moans continually and so loud that it may be heard some distance, loses its appetite and end, and has no action of the bowels: manure, if any, is as black as ink; if a fresh cow, the milk dries up entirely within three hours, and the animal almost invariably dies within forty-eight hours. On opening the animal blood is found in bunches in the veins, the flesh is bloodshotten on the stomach, and inflammation of the bowels is revealed. No very close examination has been made here. Various remedies have been tried, and I have succeeded in curing one of my own cows that was attacked by the disease. I gave her one pound of Glauber's salts dissolved in warm water, and every hour for six hours gave her a quart of strong boneset tea, rubbed her body and joints with a woolen rag to start the circulation of the blood, and in a week after the attack she was able to raise her end, but gave no milk for two weeks. At the end of three weeks she appeared as well as ever, and is all right now. This disease prevails on sandy soil, where the feed is good and the water is pure. It is new to us, and is alarming. No cattle have been attacked with it since they were taken from pasture and shut up.

Mr. JOHN ARMSTRONG, Coryell, Coryell County, Texas, says:

Having resided on a farm in this county for twenty-two years, and knowing something of diseases among horses here, I will try to answer some of your inquiries. Spanish fever, when I first came here, was the dreaded disease, but I think as soon as horses are acclimated they are less subject to it, and it is also less fatal. Before the war, I lost several valuable animals by it. Symptoms: Moping around, or standing still much in one place; very high fever; slightly swollen in the throat; great difficulty in swallowing; inability to lower the head to drink: stiffness in the hind parts and tenderness in the loins; a slight bran-and-water-looking discharge from the nostrils. The duration of the disease, which generally terminates in death, is from four to five days, sometimes the animal lingers for several days.

The first animal of mine that recovered was a large Tennessee mare, twelve years old and in fine condition. As soon as I observed the first symptoms of the disease I bled her copiously, and in three hours after she could drink water from a bucket by holding it up to her. In about five hours she ate a wheat-bran mash (one gallon), an in twelve hours had a fine appetite, eating and drinking all I would give her. She was well, but weak from the loss of much blood. She was never sick afterward, and died in colting, at the age of nineteen. Of the second case, a wild, unbroken four-year-old gelding. I bled him till he staggered, put water up for him in a trough, and sheaf-oats, and left him loose in the lot, as he was too wild to drink from a bucket held by a man. He recovered at once from the disease, but, like the mare, shed off his hair until his back and sides were naked. Since then I have lost none. Seeing the remedy at page 39 of the Agricultural Report for 1869, I have used it with entire success, greatly preferring it to bleeding, which weakens so much that the animal is unfit for service for some time after. Many here believe the Spanish fever and the so-called epizootic to be the same. However that may be, animals having green, nutritious grasses, or a green wheat field to run on, will not have either disease to hurt them.

Mr. W. E. GRANT, Carrollton, Carroll County, Kentucky, says:

We are troubled more in this immediate locality with the loss of hogs than any other class of farm-animals, and my observations have been confined chiefly to the progress of the disease called hog-cholera, and as it relates more nearly to young pigs from four to twelve weeks old. Among the first symptoms are shivering, slow and careful movements, and a desire to remain almost constantly in the warmest sleeping place they can find. They eat very little. In those that are not weaned, and in some that have been, a thick wax collects on the eyelashes and fastens the lids together. On opening the lids by force the ball of the eye appears perfectly white, and is entirely devoid of sight. The discharge from the bowels at first is like thin wheat-flour dough, but toward the latter stages of the disease becomes quite black, and has a very offensive odor. Coughing is very frequent—often one of the first symptoms. The attack lasts from five to ten days, sometimes longer. Should any apparently recover they rarely ever become of any value.

No remedies have proved beneficial to young pigs, though many have been tried. If the brood-sows were kept in perfect health the pigs most likely would not be attacked. The most successful treatment for preventing the spread of the disease that has been tried here is as follows: Remove all affected ones from the drove as soon as the first symptoms are observed. They had better be killed and buried, but may be put in a remote lot by themselves. Change the diet of the well hogs as much as possible; keep by them at all times a mixture of coal-ashes (seven parts) ground sulphur (two parts) and one part of pulverized copperas. All the coal-ashes and fine coal that the hogs will eat should be given to them.

With all the light we have on the subject we are still very much in the dark, and some farmers have become so much discouraged in their fruitless efforts to arrest the disease when it once gets among their hogs that they have given up swine-raising in disgust.

Chicken-cholera has given much trouble to poultry-raisers here lately. The most noticeable symptoms are drowsiness, disposition to remain all day on the roost, and an active discharge from the bowels. A great many remedies have been used, but none have proved of any permanent benefit. Five drops of carbolic acid in a half gallon of water for the fowls to drink seems to have arrested the disease for a time in some poultry-yards.

Mr. L. S. Morrow, Duvall's Bluff, Prairie County, Arkansas, says:

Cattle have been subject to two diseases here, both of them showing symptoms similar to dry-murrain. Various remedies have been used, but with very little success. Some say that turpentine, used internally and externally, is a good remedy, but I know of but very few cases where any benefit was derived from its use. Home-made lye-soap has been used in a few cases with slight success.

Hogs have died largely of a disease called cholera by many farmers, but by examining those that died the trouble was found to be caused by worms about three-fourths of an inch in length. These worms were found in great abundance in all the hogs examined.

Chicken-cholera has prevailed here for the past three or four years, and many fowls have died during that time. The symptoms are about the same as elsewhere.

Mr. M. J. Saddler, Dexter, Stoddard County, Missouri, says:

Hogs are seriously afflicted here with a disease called cholera. When affected with this disease they appear mopish, refuse to eat, have high fever, break out in lumps, bleed at the nose, and die soon. I have known a few cases where poke-root tea affected a cure, but the best remedy is equal parts of logwood and blue vitriol, steeped and administered to the animal.

Mr. E. T. Bently, Tioga, Tioga County, Pennsylvania, says:

The hog-cholera is unknown in our county. The only disease affecting hogs in this county is an old malady, known as throat distemper. Sulphur and ashes placed in the trough where they eat, once a month, is a preventive, but not one farmer in twenty takes this precaution.

Mr. Horace Martin, Corning, Holt County, Missouri, says:

I have been a resident here nine years, and during that time no disease has prevailed among farm-stock, except a disease among swine. Raising corn and feeding cattle and hogs is the principal industry in this vicinity. During the last three years the losses among hogs have been greater than heretofore within the circle of my observation. There is a singularity about the spread of the disease which to me is unaccountable. Some years a farmer will lose nearly his entire stock, while his neighbor adjacent will remain entirely exempt from it. Then in a year or two the conditions will be reversed. I will give you the statistics of the last three years of this and adjacent sections, numbering the farms 1, 2, 3, &c., my locality being farm No. 1. Two years ago at this date (September 1) N. Rosalins, No. 9, lost 210 head of hogs out of his feeding pens. He did not count the young shoats, which were not confined. No. 8 lost 30, but in a tenant's pen at his own feeding-yard he lost over 200 head more. Farm No. 5 lost 36; others none. Last fall No. 1 lost 57; No. 2, 80; No. 3, 45; No. 4, 64; No. 6, between 30 and 40; others none, or very few. This fall, in a herd of 150, I have lost none. Neither has 2, 3, 4, and 5, while 6, 7, and 10 have lost over 100 head. These were fattening-hogs, not shoats, and weighed from 200 to 350 pounds.

The characteristics of the disease are various, although in numerous cases no symptoms of disease were observable. In the morning I would find hogs dead that the night before I thought were well; yet on examination I would find the lungs, intestines, and skin very red and engorged with blood, but I supposed it was a natural consequence of their dying with all their blood in them. Unless the hogs are quite young the liver is always found ulcerated and otherwise diseased. The first symptom noticed is reluctance to leave their beds. Rout them out and they walk as though they were stiff. Their urine is highly colored or bloody. Possibly they may bleed at the nose; then they are sure to die in less than twenty-four hours. When found dead the nose is nearly always bloody. Sometimes the disease commences with a cough, panting at the sides and flanks, and a refusal to eat. They then linger along for a week or ten days, when they usually die. With the experience I have had with it I believe it to be more properly a typhoid fever. There are numerous remedies for sale, held as secrets, yet I never see any good effects produced by them when used in a herd of sick animals. They may be valuable as preventives. For eight years past I have endeavored to keep up my herd to one hundred and fifty head, feeding from seventy-five to eighty each year. Except during last fall I have had no disease among them. I have dry, open sheds for them to sleep in, and feed them all the ashes we make, mixed with a little salt. Occasionally we mix several tablespoonsful of sulphur, or about half as much copperas, with the ashes, say once or twice a month. We give them all the corn they will eat up clean.

Many hogs are dying all through this section of the country. I think you have undertaken a difficult task in trying to convince the average Congressman of the necessity and utility of appointing a commission to investigate the diseases prevalent among farm-stock. But it is a subject that calls for immediate investigation. This immediate vicinity has not, I think, suffered larger losses in comparison with the number of hogs kept than other neighborhoods in general. Yet this school district, comprising four full sections of land and two fractional sections bordering on the Missouri River, has in the last three years lost certainly ten thousand dollars' worth of hogs. Hence the aggregate losses in the State must reach high in the millions of dollars.

Mr. W. C. HAMPTON, Mount Victory, Hardin County, Ohio, says:

The disease among hogs does not seem to be so fatal in our county as in many other places. From the result of investigations I should say the disease was intestinal fever, or perhaps consumption. The first symptom of the complaint is a bad cough and a refusal to take food, especially corn in the ear, which they will smell of and pass by. Perhaps their jaws are too weak to crack the grain, for they will eat it when ground into meal. They continue to lose flesh for a month or more, when they die. A few have so far recovered as to permit fattening. Upon examination the livers and lungs of these animals are found greatly deranged, both being covered with white spots. Another peculiarity is that the intestines and stomach are very much reduced in size, which I think would indicate the effects of a high state of fever. No remedies have proved of any benefit. We have tried sulphur, tar, and copperas. Those saved were fed freely on corn-meal. This may have had a good effect in keeping up the strength of the animal until the disease abated or was worn out.

Chicken-cholera has been severe in some sections of this county. In this locality it was more modified and slow, but finally sure in its operations. They would mope around for weeks before death ensued. The disease must be much the same as that which afflict hogs, as the liver is found greatly enlarged and in a decaying condition.

Mr. GEORGE W. PARKER, Vandalia, Audrain County, Missouri, says:

There are but few fatal diseases among cattle here. Sometimes they die of a disease called black-leg. I know of no remedy, but salt given at regular intervals at all seasons of the year will be found a good preventive. I have handled many cattle at a time, and with very good success. In a herd on the grass beside mine twelve or fourteen head were lost while I lost none. I salted my herd regularly, while my neighbor failed to take this precaution. I salt twice a week, and have [regular days for so doing.

The prevailing disease among hogs is called cholera, but I have my doubts about its being that disease. They are attacked in a great many different ways. Some die suddenly and others linger for a long time. Breeding young hogs has a tendency to produce diseases. A general preventive will be found in breeding from none but old and mature animals.

Mr. H. M. ENGLE, Marietta, Lancaster County, Pennsylvania, says:

My own immediate vicinity has thus far been almost entirely exempt from epidemic diseases among farm-stock, except epizootic among horses several years ago and chicken-cholera now and then. The former made a clean sweep, i. e., few animals escaped the disease. The cause or causes I have never had satisfactorily explained. Chicken-cholera I had for the first time last summer, and I am confident it originated in neglecting to keep their roosting places regularly cleaned. I have no faith in any of the nostrums so generally recommended, but have in pure air, pure water, a change of feed, and a clean feeding-place. The cause is a disordered condition of the bowels, similar to that of cholera in the human; and anything that will restore them to their normal condition will effect a cure. By attending to sanitary requirements, and feeding whole grain well dried, even to browning, effected a speedy cure in my fowls.

Mr. J. S. N. Newmyer, Lone Lake, Mason County, Missouri, says:

Heavy losses have been sustained in this section by a disease among swine called cholera. There are several different diseases classed under this name, or else the disease has many different phases. I have been raising and fattening on an average about 100 hogs per year, and had very good luck until last January, when my animals commenced dying, and since then I have lost 200 head. I had 120 head on the first visitation of the disease, and out of that number lost 100. I did nothing to prevent the spread or to cure the disease—only separated the well from the sick hogs, but this seemed to do no good. Those first attacked died in a few days, and were full and plump when death ensued. After a few weeks they lingered along for a good while, and were generally reduced almost to skeletons before they died. The last ones that died had what we here call thumps, and they lingered along three weeks before they died. A very few recovered after they had become so poor and thin that they could scarcely stand. At the expiration of about two months I commenced buying another herd, weighing from 80 to 140 pounds each. I purchased them at different points, getting from six to ten at a place. After a little while they also commenced to get sick and die, and I lost 15 out of 46 in that lot. I used remedies with this herd, but do not think with any good results, although several of them recovered. Some of them had high fever, and others passed bloody urine. This was in May and June. In September following my pigs took sick, and in a very short time I lost 35 out of a herd of 38. These pigs were suckling at the time the disease broke out among them. They and their mothers were confined in the same lot with 40 hogs I was fattening, but none but the pigs were affected in any way. I have known several such cases in this neighborhood, and therefore I am not inclined to believe that the disease is contagious.

I believe an investigation, as you propose, will result in much good. It should be made thorough and complete, and the disease is so wide-spread and involves such vast interests that the government should afford ample means to investigate, and, if possible, determine its cause or causes.

Mrs. J. S. Yost, Pottstown, Montgomery County, Pennsylvania, says:

We have had some cases of pleuro-pneumonia among horses in this section of the county. Symptoms: The animals lag in their walk, and manifest little desire for food. They have a cough, with discharges at the nose and mouth. The remedy used is forty drops of aconite and eighty drops of muriate tincture of iron in water, given twice a day. The animal should be well rubbed. I am informed by a veterinary surgeon that horses afflicted with the epizootic five years ago are more liable to this disease than others. The disease is quite fatal. Some horses live but a few days, while others may linger for several weeks. If proper remedies are immediately used two-thirds will recover. A *post-mortem* examination reveals the pleura in a high state of inflammation, presenting a purple-red color. The blood is watery, and about the lungs is found pus.

A few cattle have also had pleuro-pneumonia. The symptoms are about the same as in horses, with the exception that the cough is harsher. Twenty-five drops of aconite and fifty drops of muriate tincture of iron in water, given three times a day, is the remedy used.

There have been some cases of hog-cholera in this locality. When attacked the animals swell and turn purple about the jowls, and have a white appearance about the nose and mouth. If not immediately attended to they will die in three or four days. Aconite in water (twenty drops) is used as a remedy. If the hog does not vomit within two hours, ten drops more should be given. Rub the neck and jowls twice with an ointment made of four ounces of iodine mixed with one pint of lard.

Chicken-cholera has prevailed here for several years past. They often die before you are aware that there is anything the matter with them. When attacked they refuse food, the comb becomes very dark, almost black, as does the flesh after death.

Mr. J. F. Tubb, Poplar Bluff, Butler County, Missouri, says:

A disease prevailed among hogs in this locality last summer, and about one-tenth of this class of farm-animals died of it. The disease was called measles. The animals would first break out in small red spots, which would soon turn into large sores. After this death soon ensued. No remedy was found that proved of any benefit.

Mr. E. Burket, Arch Springs, Blair County, Pennsylvania, says:

There have been very fatal diseases prevailing among fowls in this locality for some years past. I have known many flocks to nearly all die, and some of them were composed of perhaps one hundred and fifty head. During the fall I lost ninety head myself. The disease seems to prevail at any season. We have found cayenne pepper, mixed with corn-meal, about as good a remedy as any. There are many other remedies used, such as alum-water, asafetida, red pepper, &c. There was no disease among fowls here until the foreign varieties were brought into this locality.

Dr. R. J. SPURR, Lexington, Fayette County, Kentucky, says:

Ten or twelve years ago a committee of physicians was appointed by the Farmers' Club of this county to investigate the subject of so-called hog-cholera, then and now very prevalent here. The undersigned was one of this committee, and during the progress of this investigation a large number of *post-mortem* examinations were made, the subjects for examination being taken at all stages of the disease, from its incipient stage to its close in death. Copious notes were made of everything observed, but through the death of the chairman they have been lost, yet sufficient facts were impressed on the writer's mind to warrant him in bringing them to your attention. This malady among hogs is so well known that a description of its symptoms and progress is unnecessary. Suffice it to say that, whatever may be its cause, it does not occur in single cases, but when a herd of hogs is attacked by it but few escape. Pigs and small shoats seem more liable to it than older hogs. It also proves more destructive to the former than to the latter. As "there is nothing in a name," this disease had just as well be known by its popular name of "hog-cholera" as any other, although the name in many cases leads to doubt and hesitancy from the fact that looseness of the bowels is expected, when directly the opposite may exist. Purging may be present in one case, and constipation in another. In the *post-mortem* examinations made it was found that the lesions of the different organs were not uniform. The liver in one case would be found engorged or inflamed, and in another not affected. In another case the stomach would be found ulcerated or inflamed, while in still another it would be found in its normal condition. Some would have inflammation of the bowels, and others not; worms would be found in the bowels of some, while none would be found in others. There was one organ, however, in which the distinctive process was very uniform; indeed, in the forty or fifty cases examined I do not remember of a single exception. This was in the lungs, and is known as inter-lobular inflammation, and incident to the early stages of the disease. In more progressed cases there was no general diffused inflammation or hepatization. There was one other thing uniform in every case, and this was in the condition of the blood. This was placed under a microscope of rather feeble power, and the blood-disks or red globules were found to be changed from their normal configuration. In recent cases the number of disks found to be changed were limited, but very general in those where death had resulted from the disease. The blood-disk in the hog in its normal condition is nearly circular, has smooth edges, and when piled one upon another resemble somewhat small heaps of silver money without the milling around the edges. The change which had occurred was a shriveling or corrugation of the edges. Their appearance brought to my mind the scalloped edge of the bush-squash of our summer gardens. The cause of this we were unable to determine, from the fact that our microscope did not possess sufficient power for the purpose. We drew the conclusion, however, that they had been pierced or penetrated by some low order of organized life which we had not the facilities for detecting.

The writer is a farmer, and raises a considerable number of hogs annually, but he has not had this disease among his swine since the investigation detailed above, although it has prevailed to a considerable extent upon adjoining farms, and in a few instances diseased hogs of the neighborhood have mixed with his herd. He has persistently pursued a course of prevention, which may or may not have been the cause of his exemption. His course has been to give his hogs salt and sulphur once a week in the proportion of two of the former to one of the latter, always giving them as much as they will eat. They should have it both in summer and in winter, and without any regard to weather. In addition to this he uses wood-ashes freely, upon piles of which he throws salt. He has pursued this course with the hope of preventing the disease, as sulphur is destructive to low orders of animal and vegetable life.

Mr. J. P. TYLER, Smithport, McKean County, Pennsylvania, says:

The most prevalent and fatal disease to which any class of farm-stock is subject here is black-leg among cattle. It rages only among dry stock and calves, or yearlings. When I came to this locality in 1870 I was told that the young cattle were dying off at a rapid rate with a disease that no one seemed to understand. I afterward discovered that this fatal disease was black-leg. With the exception of the past season, it has prevailed every year since. It seems to be more prevalent and fatal during the hot season of the year. I have never known a case to recover. The disease comes on suddenly, and generally terminates in death within from twelve to twenty-four hours. The symptoms are a swelling of some part or parts of the body, stiffness of the limbs, and sometimes 'short and quick breathing. A blubber appears under the skin of the part swollen, and the flesh becomes black, hence the name given the disease. The best preventive known is bleeding in the neck. Feeding of saltpeter with provender is also said to be a preventive, but is not so sure as bleeding.

Mr. JOHN D. COOKE, Wheatland, Hickory County, Missouri, says:

Hog-cholera prevails here, and I have had it to contend with. After fruitless experiments I am satisfied there is but one thing that will cure or prevent the disease, and that is too expensive. I am sure that if a beef could be slaughtered every day or two, and the carcass given to the hogs, that the well ones would escape the disease, and those not too sick would be cured. Blue-mass for fowls suffering with the cholera will be found of more real value than anything else.

Mr. E. W. HAMLIN, Bethany, Wayne County, Pennsylvania, says:

Black-leg, or hoof-disease, among cattle has proved fatal in numerous cases here. But very few of those attacked recover.

Large numbers of young pigs died last spring from pneumonia or inflammation of the lungs. No remedy was found.

Mr. J. T. HESTER, Corsicana, Navarro County, Texas, says:

Horses and cattle are measurably healthy, but hogs are pretty generally affected with something like cholera, and the losses are quite heavy. No specific remedy has yet been found, but a teacupful of turpentine to one half bucket of shelled corn, well mixed, has proved quite beneficial.

Fowls are subject to a disease which causes them to droop around for a few days and then die. They rarely recover. The disease is generally called cholera. No remedy has been found. Your department will confer a great blessing on the country if it succeeds in finding a cure for cholera in hogs and chickens, and liver-rot in sheep. This last-named disease can be prevented, but with present lights on the subject it cannot be cured.

Mr. WILLIAM S. RAND, Vanceburg, Lewis County, Kentucky, says:

Hogs being the staple product and source of the principal revenue of this county, I have given special attention to their treatment and the diseases to which they are incident. In the limestone sections of this county the fatality of diseases has been most disastrous. Hog-dealers have tried all the remedies and practiced every kind of treatment. In herds where an animal has died those remaining have been separated and quartered in small lots in distant localities, and this treatment has generally been more successful than any other. The symptoms of the disease are widely different, and what will cure one would seem to kill two. Sometimes temporary relief may be obtained, and the animal apparently be in a fair way of recovery; but in all probability in a day or two afterward it would be found in a dying condition. Mr. Brazil Lyle, a hog-raiser in the mountains, has been successful in treating the disease with the free use of coal-oil, given in half pints and by injection. The same remedy has failed elsewhere. Capt. Jack Henderson, who has had large experience in the treatment of the disease, has arrived at the conclusion that it is incurable. He has tried all the remedies, but his losses have been very heavy.

It has been stated and generally credited that the mountain or mast fed hogs escape this disease. In order to satisfy myself on this point I this fall made a protracted trip to the mountains of Eastern Kentucky for the purpose of observing the operations of the disease in the very highest latitudes of the State. In two instances the whole of two herds of fat hogs, ready for the market, died within two days, shortly after my arrival. They had previously shown no symptoms of the disease. Other lots, in the same neighborhood, showed no signs of disease.

It is most painful to witness the disastrous results of this mysterious and fatal disease on the young farmers of the interior. They grow a crop of corn to feed to hogs, buy the hogs generally on credit for a few months, and then, when they are almost ready for the market, this scourge comes along and carries them all off. The farmer is left without corn or other supplies for his family, and is also in debt for the hogs which he has lost. I could name several instances where the wolf is now at the door of many of the hard-working, honest farmers of this section, and if it is within the means of your department and the agency of the National Congress, in the name of God and humanity push forward the work for the speedy relief of the great producers of the land.

Fowls are subject to sudden and fatal attacks of disease. I know many farmers who have lost hundreds of fowls without warning. At the present writing the disease is prevailing extensively. All the remedies and preventives known have been used without effect. Whatever the papers publish is greedily accepted and tried, but generally what succeeds in one locality fails in another. Those who take the most pains to avoid the disease suffer the most. New locations and fresh walks, with pure water to drink, is highly recommended. The latest preventive and cure is tobacco-pills, given when the fowls are drooping, either in food or otherwise.

Mr. L. D. VAN DYKE, Clarksville, Red River County, Texas, says:

The most serious disease affecting horses and mules and causing serious loss among these animals is called "blind staggers." For many years it was the prevailing opinion here that this disease was caused by feeding worm-eaten or unsound corn, but in 1875 our loss in this and adjoining counties was very serious, and we never had heavier or better corn than in that year. Stock not in use and fed entirely on grass are not liable to contract the disease, and nearly all those attacked would recover if they were turned on grass as soon as the first symptoms appear.

The first symptom of the disease is a disposition to sleep, and a dull, stupid appearance generally. As the disease progresses the animal becomes blind, and the disease soon assumes the form of brain-fever. Some die in twenty-four hours, while others may linger for weeks. I have relieved several by boring through the skull to the brain with a small penknife. They recover their sight immediately and become very docile; but it is evidently a disease of the stomach, and I think much of it is caused by too severe labor when the stomach is full, although it has raged here as an epidemic.

A disease affected the hogs in this county last spring which caused very great loss. It was evidently a disease of the lungs, as the symptoms were a dry cough and difficulty in breathing, similar to that produced by eating cotton-seeds. About 50 per cent. of those affected died. I am told that all were cured that were given copperas and gunpowder. Here in the South, where for many months in the year hogs have to find their own support as best they may, I attribute most of the diseases to which they are incident to worms. All diseases to which they are subject here are called cholera; but I have had no experience with regular cholera in hogs.

Much of the disease among chickens called cholera is produced by their eating henbane or nightshade, which grows very plentifully throughout this country. The chickens eat it with avidity in the spring season.

Mr. R. S. BROWN, Bethlehem, Northampton County, Pennsylvania, says:

Since the epizootic malady some years ago among horses, the most fatal disease, and the worst one that we ever experienced, was the spinal disease among the same class of animals. The horse would be taken out of the stable apparently well, and after being driven a mile or two would fall down, completely paralyzed, and unable to get up behind. As this disease appeared in the early part of winter, during snow and bad roads, unless within calling distance and with the assistance of a dozen strong men, the animal was in danger of perishing on the spot. I have taken mine home on low sleds, rolled them off into large and warm stables, padded them all round with straw to keep them from knocking their brains out in their frantic efforts to get up, and then used the following remedies with success: I took an empty salt-bag and filled it with clover-heads. Upon this scalding water was poured, and it was then applied to the skin as hot as could be borne. This was renewed every half hour by careful men during the whole night. The horse was then rubbed dry and a mild laxative medicine used for a few days. After that the horse was raised to his feet by means of a side of leather to which was attached rings and pulleys. This was done at intervals of six hours. The horse was allowed to stand about thirty minutes, when he was let down again, and this operation was then suspended for eight or ten days. If left to lie without being compelled to stand up they will never recover. Veterinary surgeons, who tried the old remedies of bleeding and purging, and applying turpentine to the spine, lost every horse so treated.

Disease among fowls has been general, and the losses have been very heavy. Neither the poor man's dozen nor the rich man's hundreds were spared. They died by hundreds and thousands. Most of them would droop for a few days and then die. Others would die upon their roosts or nests. Young and old seemed to fare alike. No remedies proved effective. The disease continued over a period of five or six years, until it was thought none would be saved. People continue to lose some, but the disease is now abating. The best remedy I know of is wheat-bran made into a thick paste with milk and liberally sprinkled with red pepper. They eat it ravenously.

Mr. AMOS WOODLING, Beach City, Stark County, Ohio, says:

Heavy losses have been sustained among fowls by the ravages of a disease called chicken-cholera. Entire flocks have been destroyed by it. The fowl becomes stupid, loses its general brightness about the head, diarrhea sets in, and the result is death within three or four days. After death the liver is found to be of a light clay color.

Dr. WILLIAM GUTCH, Blakesburg, Wapello County, Iowa, says :

I have special inquiry with reference to the disease among swine known here as hog-cholera. The symptoms are—the animal refuses to eat, vomiting and purging set in occasionally, but not uniformly ; there are unscular twitchings of various parts of the body. Death often takes place within twenty-four hours. The causes of this disease are very obscure, as it occurs under every variety of sanitary condition. But from its extensive prevalence here, and from the manner in which it frequently spreads, I have no doubt but that it is highly contagious. Nearly every kind of treatment (including the use of many vaunted specifics) has been tried by the farmers here, but the uniform testimony is that they all fail to produce any benefit.

For several years past a highly fatal disease has prevailed in this section of country among different members of the gallinaceous family, viz., turkeys, Guinea hens, and domestic fowls. It is known here as chicken-cholera, and occurs especially where great numbers of fowls are kept together, and is, I believe, caused by bad sanitary conditions. The animal mopes around, has an uncertain gait, purges, and usually dies in a very short time—sometimes in a few hours, but generally not for several days. The remedies are thorough cleanliness and ventilation, with lime, sand, and a mixture of corn-meal and Venetian red. These, if they do not cure the disease, will usually prevent its further spread.

Mr. J. W. MEANS, Carthage, Jasper County, Missouri, says:

From my own experience I am prepared to say that hog-cholera is contracted by feeding and watering them in unclean places. My neighbor is now losing hogs every day, while I am losing none. I have about two hundred head on an adjoining farm. His hogs are fed in unclean places and sleep about old straw stacks, and mine do not. I feed plenty of salt, lime, and soda, and am satisfied hogs will not take the disease if given these preventives in time. I feed hundreds of head, and have never lost any.

I also feed a great many cattle, but have never lost any by disease. I use salt, ashes, and madder twice a week. A great many people lose cattle by allowing their hogs to run with them. When eating green corn the hogs chew the stalk until all the strength is extracted. They then drop it; it dries; the cattle eat it, and it clogs the stomach and produces what is called the "mad itch." A great many cattle have died from this disease this year.

Black-oak bark boiled in water to a strong sirup, and kept where the fowls can have access to it, will be found both a remedy and preventive for chicken-cholera. Sulphur fed with corn-meal is also good. Lime and ashes will destroy the lice which infests them.

Mr. F. M. CUMMING, Harrisonville, Cass County, Missouri, says:

This immediate vicinity has suffered immensely from the ravages of the disease known as hog-cholera. During the early part, and indeed almost entire winter, it disappears or scarcely makes an appearance, but during the months of March and September it breaks forth in most fatal forms, frequently causing the death of every hog on a farm. I have known as many as sixty and eighty to thus die in one week on a single farm, leaving not one to commence restocking with. The disease assumes two forms. The first, and what I presume to be cholera proper, commences with black discharges from the bowels, which continue until the animal "wears it out," or becomes a gaunt skeleton and dies from mere exhaustion. The second form commences with an utter refusal to eat, stupid appearance of the animal, high fever, very constipated bowels, and a great desire for cold water. This form generally proves fatal in three or four days. When dead, blood gushes from the nostrils, and upon examination the lights resemble coagulated blood of the consistency of cream. This disease causes greater losses to the farmers of this corn-producing country than all other diseases affecting farm-animals combined.

Mr. W. B. HARSHA, Harshasville, Adams County, Ohio, says:

A disease prevails among hogs in this section which is generally called cholera. Several herds were attacked by it during the dry weather of last fall. But I think instead of its being cholera it was pneumonia. The first noticeable symptom was coughing. Then follows fever, no desire for solid food, and constipation of the bowels. About all those first attacked died, but after we commenced doctoring we saved some. We gave a physic of sulphur and saltpeter, followed by the use of Fout's cattle-powders, or fluid extract of aconite and belladonna, equal parts, and one teaspoonful at a dose. Toward the last this cured nearly all the animals affected, and we believe the use of the cattle-powders prevented the further spread of the disease.

Mr. THOMAS D. TYLER, Ottumwa, Wapello County, Iowa, says:

Hog-cholera has been the prevailing disease among farm-animals here, and it has been very fatal and destructive. There is more or less of this disease every year, and those attacked seldom ever recover. The cause, in my opinion, is the lack of proper care. Large numbers are kept together, and they are allowed to sleep in old rotten straw-stacks, which engenders disease. With proper care I think the disease would entirely disappear.

Until this fall cattle have been very healthy in this vicinity. A disease is now prevailing among them which is very destructive. It is called "black-leg." The Weekly Courier of this place says: "William Shepherd, living four miles north of this city, has within a very short time lost eleven head of fat cattle from a disease which seemed to baffle the skill of the most successful veterinary surgeons. Yesterday he lost another fine animal, and sent for Dr. Hinsey for the purpose of holding a *post-mortem* examination. The examination was held, and the doctor informs us that he found that the cattle had been dying of 'malignant anthrax,' or black-leg. In the case he examined he found the cavities of the heart occupied by a clot of blood as black as ink, and nearly the size of his fist. The mass was firm and tough, and when removed the blood of the arteries, of the same consistency of that of the center of the heart, followed its removal in strings the full size of the arteries and several inches in length. The disease is very contagious from a dead carcass, or from the blood of the animal when tasted by other cattle. * * * The doctor gives it as his opinion that when stock is affected by this disease the farmer would, in the event the case proves fatal, do well to bury the carcass of the animal without even removing the hide. So far as known there is no remedy for the disease, and the best thing to do is to prevent its spread. Two or three other farmers have recently lost cattle."

The following treatise on this disease is from the pen of Professor Shaefer:

"Its attacks are confined almost entirely to animals that are in high condition or rapidly improving; we should say too high condition and too rapidly improving. In some instances the disease will give some warning of its approach; but generally the beast will appear to-day perfectly well and to-morrow he will be found with his head extended, his flanks heaving, his breath hot, his eyes protruding, his muzzle dry, his pulse quick and hard—every symptom, in short, of the highest state of fever. He utters a low and distressing moaning; he is already half unconscious; he will stand for hours together motionless, or if he moves or is compelled to move, there is a peculiar staggering referable to the hind limbs, and generally one of them more than the other; by and by he gets uneasy; he shifts his weight from foot to foot; he paws faintly and then lies down; he rises, but almost immediately drops again; he now begins to be, or has already been, nearly unconscious of surrounding objects.

"There are many other symptoms from which the different names of the disease arose. On the back or loins or over one of the quarters there is more or less swelling. If felt when it first appears it is hot and tender and firm; but it soon begins to yield to the touch, and gives a singular crackling noise when pressed upon. One of the limbs likewise enlarges, sometimes through its whole extent, and that enormously. It, too, is at first firm and hot and tender, but it soon afterward becomes soft and flabby, or pits when pressed upon, *i. e.*, the indentations of the fingers remain. When examined after death, that limb is full of red putrid fluid; it is mortified, and seems to have been putrefying almost during the life of the beast. Large ulcers break out in this limb, and sometimes in other parts of the body, and almost immediately become gangrenous; pieces of several pounds in weight have sloughed away; three-fourths of the udder have dropped off, or have been so gangrenous that it was necessary to remove them, and the animal has been one mass of ulceration. The breath stinks horribly; a very offensive and sometimes purulent and bloody fluid runs from the mouth; the urine is high-colored or bloody, and the fæces are also streaked with blood, and the smell from them is scarcely supportable.

"In this state the beast will sometimes continue two or three days, at other times he will die in less than twelve hours from the first attack. In a few instances, however, and when the disease has been early and properly treated, all these dreadful symptoms gradually disappear, and the animal recovers.

"It is to a redundancy or overflowing of the blood, the consequence of the sudden change from bad to good living, that this disease most commonly owes its origin. It is most prevalent in the latter part of the spring and in the autumn, and very often at these seasons of the year proves destructive to great numbers of young cattle in different parts of the States. It is sometimes, however, seen in the winter and the early part of spring, when the cattle are feeding on turnips. Some situations are more subject to this complaint than others. It is most frequent in low, marshy grounds and pastures situated by the side of woods.

"It is a disorder of high condition and over-feeding. The times of the year and the character of the cattle prove this. It occurs in the latter part of the spring, when the grass is most luxuriant and nutritive, and in the autumn, when we have the second flush of grass; and the animals attacked are those principally that are undergoing the process of fattening, and that have somewhat too suddenly been removed from scanty pasturage and low feeding to a profusion of herbage, and that of a nutritious and stimulating kind. The disease sometimes occurs when the cattle have been removed from one pasturage to another on the same farm; but more so when they have been brought from poor land at a distance to a richer soil. There are in the latter case two preparatory causes—the previous poverty and the fatigue and exhaustion of the journey.

"This disease rarely admits of cure, but fortunately it may in general be prevented. If the malady is discovered as soon as it makes its appearance, the beast should be immediately housed, and then from four to eight quarts of blood taken away, according to the age and size. Two hours after bleeding give a purging drink, as follows: Epsom salts, 1 pound; powdered caraway seeds, ¼ ounce; dissolve in a quart of warm gruel, and give (which will be found of a proper strength for young cattle from the age of one to two years).

"The bleeding should be repeated in three or four hours, if the animal is not materially relieved; and a third bleeding must follow the second, if the fever is unabated. There must be no child's play here; the disease must be knocked down at once or it will inevitably destroy the beast. The physic likewise must be repeated until it has had its full effect.

"As soon as the bowels are well opened the fever-drink No. 1 (tartar-emetic, 1 drachm: powdered digitalis, ½ drachm; niter, 3 drachms; mixed in a quart of thick gruel) should be administered, and repeated morning, noon, and night, all food except a little mash being removed.

"At the first appearance of the disease the part principally affected should be fomented several times in the course of the day with hot water, and for at least an hour each time. For this purpose there should be two or three large pieces of flannel in the water, that after one of them has been applied thoroughly hot and dripping to the part affected, another equally hot may be ready when this gets cold.

"As soon as the fever begins evidently to subside and the beast is more himself and eats a little, the fever-medicine must not be pushed too far. It should be remembered that this is a case of highly inflammatory disease which soon passes over and is often succeeded by debility almost as dangerous as the fever. The ox therefore must not be too much lowered; but, the fever abating, mildest tonic drink (gentian, 2 drachms; emetic tartar, ½ drachm; niter, ¼ ounce). Give in gruel.

"If this does not bring back the fever it may be safely continued once every day until the ox is well, or the quantities of the gentian may be increased and the emetic tartar lessened and at length altogether omitted, the niter being still retained.

"A seton (of black hellebore root if it can be procured) should be inserted into the dewlap, and if the beast can be moved it should be driven to much scantier pasture.

"Should not the disease be discovered until there is considerable swelling and a cracking noise in some tumefied part, a cure is seldom effected. Bleeding at this stage of the complaint can seldom be resorted to, or at least one moderate bleeding only should be practiced, in order to subdue any lurking fever that may remain. If a cure is in these cases attempted, the tonic drink should be given, which may invigorate the system by its cordial and tonic powers, and prevent the mortification extending.

"The swelled parts should be frequently bathed with equal portions of vinegar and spirits of wine, made as hot as the hand will bear; or if ulceration seems to be approaching, slight incisions should be effected along the whole extent of the swelling, and the part bathed with spirits of turpentine, made hot.

"If ulceration has commenced, accompanied by the peculiar fetor that attends the disease, the wounds should be first bathed with the disinfectant lotion (solution, chloride of lime in powder, ¼ ounce; water, 1 pint; mix).

"The hot spirits of turpentine should be applied immediately after this and continue in use until either the mortified parts have sloughed off or the sore begins to have a healthy appearance. The tincture of aloes or Friar's balsam may then follow.

"Since so little can be done in the way of cure, we next anxiously inquire whether there is any mode of prevention. The account which we have given of the disease immediately suggests the prevention, namely, to beware of these sudden changes of pasture; now and then to take a little blood from, or to give a dose of physic to, those beasts that are thriving unusually rapidly, and whenever the disease breaks out on the farm to bleed and to purge and remove to shorter and scantier feed every animal that has been exposed to the same exciting causes with those that have been attacked. The farmer should be particularly watchful during the latter part of the spring and the beginning of the autumn. He may thus save many a beast, and the bleeding and the physic will not arrest but rather assist their improvement. He who will not attend to a simple rule like this deserves the loss that he may experience."

Mr. James C. Fairbank, Concord, Morgan County, Illinois, says:

In cattle some heavy losses have been sustained from "Texas fever," so called. The disease has been confined mostly to native cattle in this vicinity, and to these only in cases where they have been in the same pasture, lot, or cars, or across the track of the Texas cattle. It does not seem to be contagious from being near in separate pastures. About two weeks after exposure the cattle cease to eat and soon die. In one case a man had eighty head of extra fine cattle, just ready to ship. In August he bought a lot of Texas cattle and turned them into his pastures. He then changed them, putting the native fat ones in the pasture where the others had been. They soon after commenced dying, and nothing seemed to check the disease until eight had died. They were sixteen-hundred-pound cattle. A neighbor went to Kansas and brought in thirty-five or forty head of steers weighing about twelve hundred pounds each. He did not know that they had been exposed in any way, and they could not have been except in the cars or in lots where they were temporarily quartered. Just two weeks after, and before he knew anything was wrong, two of them died. He finally lost nearly half the lot. Some effort was made to doctor them, but without success. The disease always disappears with heavy frosts.

Several cows have recently died, just after calving, with milk-fever. The only thing I have known to help them was to drench freely with melted lard and turpentine; say one pint of lard to two tablespoonfuls of turpentine, and repeat the dose, if necessary.

The hog-cholera, so called, has been the greatest scourge we have ever experienced. During some years from 60 to 75 per cent. of the hogs are lost by it. The usual symptoms, as now manifested, are a loss of appetite, cough, an inclination to scratch and sometimes to thump, and general lassitude. They then incline to "pile up" in their beds, and many of them die during the night. No purging or vomiting is observed, but rather a severe constipation, and the excrement is dry and hard. Many die just after drinking, especially fat ones. Some will eat their regular amount of feed until just before death, while others will become greatly emaciated and linger for weeks before death relieves them. Mr. Thomas Danby, of the English settlement, says he had a large sow to lie three weeks without either food or water, and then get well. Some years since a few of my fat hogs cracked open on the back. These cracks extended to the bone, and in some cases the fat and flesh sloughed off. A few affected in this way recovered.

Most hogs that die of cholera will bear gathering up and hauling to the grease-factory; but a neighbor of mine, who had some very fat ones die of a sort of congestion, attempted to skin them, but they were so offensive that he had to desist. The blood had settled all through them, and had turned the fattest portions of them very black. The bones were very tender and apparently rotten.

The disease seems to have no fixed or certain symptoms. Sometimes it will only attack young pigs, and only ceases when there are none left to kill. Entire litters often die while the mother remains comparatively healthy. In other instances only fat hogs may be attacked, but generally the heaviest losses are sustained among shoats weighing from 75 to 125 pounds. Very often its sweeps over a whole neighborhood, and rages as a contagious epidemic. In such cases only those exposed to the disease suffer, while isolated herds remain exempt. Upon the first evidences of the disease it has got to be the practice to separate the hogs and scatter them over the farm as much as possible, and if they are being fed on dry food to change them to grass. This course seems to have a tendency to check the disease. The losses generally range from 25 to 80 per cent. of all the hogs; sometimes it reaches even higher than this. I made an estimate once of a circle of one mile, taking my own place as the center, and within that circle 66 per cent. died. Mr. Danby, spoken of above, lost 160 out of 200 head.

The so-called cures are various, but as cures they are mostly failures. Preventives are often used with great benefit. But, however strangely it may seem, what may be successful as a preventive or cure in one case may utterly fail in others. Mr. Danby tried turpentine, using in one season ten or twelve gallons mixed in swill, but without success. He now feels that he has found a sure remedy in the use of quick-lime, ashes, and salt. He feeds it to his hogs once or twice a week, and if they are coughing and not eating with their usual relish he keeps it constantly in their feed-troughs. Since he commenced using this preventive he has lost no hogs. Mr. Carter, a relative of mine and a large hog-raiser, says he has never been troubled with cholera to any great extent. His reliance is upon the use of turpentine, salt, and ashes, regularly and steadily given.

J. M. Thompson, a neighbor, thinks he has a certain preventive and sometimes a cure for the disease, in a mixture composed of arsenic, copperas, sulphur, asafetida, lime, salt, and ashes. He feeds to them once a week, and, if cholera is around, oftener. He regards arsenic as the main ingredient. Samuel Newton recommends to every one the use of copperas, sulphur, and black antimony. He says their constant use has proved of great benefit to him, as well as to others to whom he recommended the prescription. Mr. H. Engleback fed a large number one year on slops made from shipstuff, bran, &c., in which he constantly used soda. He had good success, while others immediately around him, who fed on corn, lost heavily. Some use ashes, sulphur, and salt, others copperas, ashes, and salt, and still another salt and ashes. These are generally used as preventives. If the hogs have cholera arsenic is given, and if they are past eating they are drenched. During one year my son utterly failed with all these articles. Nothing whatever seemed to do his animals any good. They were large, fine, fat hogs, in apparent good health, yet they died daily. After a great many died he was advised to use mutton tallow. When this was used freely it seemed to check the disease, but when he ceased to use it, because of the expense, the disease returned with great fury, and swept off from 60 to 75 per cent. of those left. He is now using J. M. Thompson's remedy, and so far with success. I know of one case where a hog, seemingly almost dead, was cured by drenching with melted lard and then giving it Mr. Thompson's arsenic mixture. The experience of all seem to be about this—preventives are a success if used regularly and judiciously, but if the hogs are once attacked there is nothing that will prove of much benefit. I think the disease is both epidemic and contagious. I have been through two sieges of it. In the first instance it was evidently epidemic, passing from east to west, and taking all in its course. In the second instance the disease was imparted to my herd by an infected shoat that found its way into my inclosure.

I forgot to mention in the proper place that I found some benefit result from burning and charring the carcasses of the dead hogs and feeding the refuse to the living ones.

This subject is one of vast importance to the farmers of this country, and I trust you may receive an appropriation sufficient to make a thorough and speedy investigation, in order that the cause may be discovered and a sure remedy be found.

Mr. H. H. MITCHELL, Lemon, Wyoming County, Pennsylvania, says:

There is no disease prevailing in this locality at present, nor has there been the past season. In an experience of over forty years as a farmer I have invariably found that an ounce of preventive is worth a pound of cure—that is, by judicious feeding and care I have found that all classes of farm-stock are less liable to be attacked by any prevailing disease than those illy cared for and in a measure left to shift for themselves. Protection from the blasts and storms of winter, plenty and frequent changes of feed, and an abundance of salt has always been my motto, and I have never lost but one cow and two or three calves by any disease in all these forty-odd years. The benefits of good care and feeding were very apparent during the prevalence of the epizootic among horses a few years since. By protection from cold currents of air, especially when the horse was wet with sweat, plenty of salt, potatoes, and laxative food generally, many animals escaped altogether, while those that did have it escaped without any serious results. I have lately heard of a very simple and sure remedy for this trouble. It is, to take five or six onions and put them in the feed box of the horse and let the animal help himself. After eating two or three he will begin to snuff and blow, when his nose will commence to run, and soon thereafter he will be a well horse again.

After having lost some valuable hens with the gapes, I took a tablespoonful of hog's fat, melted it, and poured it down the throat of one so near gone that it could not stand up. In a day or two, without other treatment, it had entirely recovered. Others may have known of this remedy. I did not.

Mr. P. E. WHITE, Denmark, Lewis County, New York, says:

A new disease made its appearance the past summer among horses, which is called spinal meningitis, and baffles all medical skill. In the last case which came under my observation the horse, to all appearance, was well in the morning but died before noon. The animal, when attacked, begins to droop very suddenly, refuses to eat, shows signs of pain, drops to the ground, and is never again able to rise. They usually lie flat on one side, and never seem to move a muscle even in the throes of death.

Colic, in its various forms, causes the death of more of our valuable horses than any other disease, and an effective remedy would be of great value to the owners of these animals. Various remedies have been prescribed for this disease, but they often fail. Colic terminates one way or the other in a very few hours, and therefore requires speedy and careful treatment.

We do not know of any prevailing disease in the herds of our county except that of abortion among cows. This direful scourge and fearful drawback to the dairying interests of this locality has been prevailing here for several years, and still continues with unabated progress. Thousands of dollars are yearly lost to the farmers by the ravages of this disease alone (we term it a disease, for we know of no other name to give it). We have known of large dairies where nearly three-fourths of the cows would abort, and yet no key has been found to unlock the mystery. It is well known that large sums of money and much time have been expended to solve this mystery without arriving at the true cause or source of the trouble. We would therefore recommend that abortion in cows be one of the diseases marked for a special and thorough investigation by your department. The welfare of the farmers and stock-growers of the country demands this.

In swine there has been more or less mortality the past season, especially among pigs from one to two months old. We have known of nine or ten pigs of that age to die one after the other, and all apparently of the same disease. The first symptom noticeable is a loss of the use of the hind parts. They commence to drag their hind legs after them, refuse to eat, and usually die within a few hours. Sometimes when affected with the disease they will give a squeal and drop dead without further ceremony. No remedies, so far as we can learn, have been administered for the arrest of the disease. Some call it a disease of the kidneys, but we are not prepared to state what it is. We only know it cleans out a pen of hogs (it prevails also among grown hogs) in double-quick time.

Mr. JOSEPH LOVE, Bacon, Coshocton County, Ohio, says:

The principal disease here among poultry is called chicken-cholera. The first thing we observe is a diarrhea. The head becomes pale and the fowl commences to droop and is disinclined to move about. There seems to be fever and thirst, the fowl drinking very often. When the internal parts are examined the liver is much swollen and is dark-colored. When a flock is attacked the greater part of them die. We have found no specific for it yet. We think the exclusive use of corn as food has a tendency to bring on the disease. In my own case I have found that rich bran mixed in dish-water and occasionally in milk makes a healthy feed; so does wheat-screenings. Pure water is very essential. Fowls are not as particular as other animals as to what they drink. They will drink drippings from manure heaps and other filthy places as greedily as from pure sources. I think cleanliness in all respects would ward off the disease.

Mr. M. STOCKING, Wahoo, Saunders County, Nebraska, says:

Previous to 1876 the swine of this county were healthy. The annual loss from disease probably did not reach 1 per cent. of the whole number. In the fall of 1876 the cholera broke out near Ashland, along Salt and Wahoo Creeks. During 1877 the disease has proven exceedingly virulent along all water-courses, and has baffled all remedies. In the beautiful valley of Wahoo fully 95 per cent. of those attacked have died. On uplands the disease has proven less virulent, many large herds having wholly escaped thus far.

The immense aggregate annual losses from disease which occur among our domestic animals, and the danger of importing others from abroad, imperatively demands national legislation and the establishment of a school where veterinary science shall be thoroughly taught.

Mr. ISAAC HOOFER, Lebanon, Lebanon County, Pennsylvania, says:

A few cases of what is generally termed cholera has occurred among hogs here, but the cases have been so few that the subject is hardly deemed worthy of notice.

Among horses the only disease deserving notice is "inflammation of the intestines," which, if not promptly attended to, generally proves fatal in a few hours. The symptoms are great restlessness, pawing the ground, letting themselves fall, and showing by many ways that they suffer great pain. The cause is chilling of the blood by drinking cold water when heated, or getting wet when heated, or sometimes it may be brought on from unwholesome food or irregular feeding. The cure is one ounce tincture of asafetida, one and one-half ounce tincture of opium, and one-half ounce of sulphuric ether, mixed with half a pint of water and given to the suffering animal.

Having made considerable inquiry among horse-dealers and horse-farriers as to the cause of diseases in horses, I feel satisfied in saying that over one-half of the diseases to which these animals are subject are brought on by irregular feeding, and three-fourths of the other half from unwholesome food and abuse.

Mr. GEORGE A. SHUMAN, Shermansdale, Perry County, Pennsylvania, says:

There has been some cases of what is called "black-leg" among young cattle in this vicinity. I don't know anything about the disease, but have heard that there is no remedy for it—that cattle that are attacked by it must die.

There have also been a few cases of cholera among hogs in this locality. There seems to be no remedy for the disease—at least all I have tried to doctor have died. Those that have free access to charcoal and mud-holes seem to escape the disease.

Chicken-cholera is very common, whole flocks dying within a few weeks. We have tried soft feed, in which alum was dissolved, and also put alum in the water which they drank, with, I think, beneficial results.

Mr. F. P. SCHOFIELD, Buffalo, Dallas County, Missouri, says:

Last year a good many hogs died here with a disease commonly called cholera. Doubtless this was the disease in most cases, but not in all. This year a few stock-hogs have died from the effects of like diseases. Few remedies have been used, and these with but poor success. Turpentine, ashes, soft soap, &c., have seemingly checked the disease in some few cases.

Mr. WILLIAM ZIMMERMAN, James X-Roads, Somerset County, Pennsylvania, says:

We are sometimes troubled with a disease known here as "black-leg" among our cattle. In most cases the animal indicates great pain, and generally dies within a few hours. If the skin be removed after death mortified spots are frequently found. I once arrested the disease, after losing half my herd, by daubing their feeding trough with pitch-tar, and feeding rosin mixed with saltpeter and sulphur.

The only trouble I ever have with hogs is the result of a kidney disease. This I generally cure by feeding corn boiled in strong lye. I also put wood ashes in the feeding-trough occasionally.

Mr. BENJAMIN M. HALL, South Eaton, Wyoming County, Pennsylvania, says:

What is called "hollow-horn" among cattle is frequent here. The remedies are to slit the tails, bore the horns, and pour peppery, irritating fluids into the ears. When this is done the animal generally recovers.

A few winters ago a disease raged among our cattle for which we had no name. They lost the use of their limbs, and would swing their heads back and forth as if in great pain and distress. They died within from six to twelve hours from the time they were taken sick. I lost five head, and I believe every animal that was attacked died. We were foddering corn-stalks at the time, and the corn-fodder that year contained an unusual quantity of smut.

What is called chicken-cholera is quite common. One-third of an ounce of calomel mixed with food for twenty full-grown fowls has been used as a remedy with good success. In neighboring towns, where fowls are kept confined in yards or pens, many are dying. The disease is called roup, for which no remedy is known here. I am not sufficiently acquainted with the disease to describe it.

Mr. J. W. STEWART, Lancaster, Schuyler County, Missouri, says:

There are no diseases that amount to anything among our farm-stock except among hogs, and they are so complicated I can scarcely describe them. Some of the animals seem to be afflicted with two or three diseases at the same time. At least four or five distinct diseases prevail in this vicinity. The first is the cholera, for which we have no remedy. The second, "thumps," which is not very fatal. Indigo-water and copperas is the best remedy so far discovered for this disease. Third, quinsy, for which no effectual remedy has been discovered. Fourth, bleeding of the nose. Fifth, enlargement of the upper jaw. In this disease the hair becomes coarse like bristles, and seems to stand on end. Sixth, a very fatal disease, which kills the animal in less than six hours. I have seen hogs afflicted with three different diseases in a pen containing but eight animals. No remedy for the three last named diseases is known here.

Messrs. M. K. PRIME & SON, Oskaloosa, Mahaska County, Iowa, say:

The breeders and pork-producers of this locality have been troubled a great deal with what is termed "hog-cholera." In pigs the first symptom of the disease is a cough. Some of them, if let run a few days or a week or two, will be attacked with the "thumps." This is the first stage of the fatal disease of cholera. The next symptoms are stupidness, loss of appetite, inclination to lie in their nests, great thirst, and continuation of cough. Some will purge freely until all nutriment seems to have passed from them. The urine becomes very red, and a slimy excrement passes from the bowels. They live but a few days after these symptoms are manifested. The symptoms of the disease are about the same in more aged and full-grown hogs. Our opinion is that the disease is caused by feeding too much rich food, and then a sudden change on to pastures. Overfeeding also produces disease. The diet of a pig when first commencing to eat, and also that of the mother while suckling, should be of light, easily-digestible food, containing sufficient nourishment to sustain them well. Should the pigs take cold and commence to cough, give them a small amount of Glauber salts, sulphur, and ginger, or something that will produce a similar effect. Farmers generally use, and with considerable success, salt, wood-ashes, soapsuds, or small quantities of soft soap.

Mr. J. A. GUNDY, Lewisburg, Union County, Pennsylvania, says:

There has been but little stock affected by any disease in this county that I have heard of. The usual disease known as chicken-cholera prevails, for which every person has his own remedy, but nothing that has proved positively satisfactory. I have found white-oak bark the best remedy. It acts as an astringent, and should be given by soaking thin feed in the liquor.

I have often had my hogs attacked with a disease which affected them in the back and legs to an extent that they could not walk. I always found them ready to eat chicken excrement, which I gave them daily in quantities of say a half spadeful. The results were always satisfactory.

Mr. E. J. HIATT, Chester Hill, Morgan County, Ohio, says:

Our time has mostly been occupied in breeding sheep. We have made examinations of flocks in Vermont, New York, Pennsylvania, West Virginia, and to a limited extent in Missouri and Massachusetts, and also in our own State. We have found that climate, soil, and lay of land have a great influence in regard to the health of different breeds of sheep. Our experience has been largely with the merino breed, but not entirely so. We consider this section as healthy and well-adapted to the successful growing and improvement of the merino as any section we have visited. In this and adjoining counties the diseases of sheep that most prevail are more destructive to young animals— lambs and tegs—from lambing until the first shearing. There is a disease prevailing here in wet seasons which is very generally called " paper skin " or " pale disease." It is probable that a number of distinct and separate diseases, or causes, are here called one disease, and given the above name. Grub in the head, tape-worm, lung-worm, stricana or strichnia, and some others are frequently spoken of by the wool-grower as one disease—" paper-skin." The lack of a sufficient quantity and proper quality of feed is a great cause or assistant to diseases.

Grub in the head is not a new disease, but it is a very difficult one to prevent or cure, and it is more or less destructive on all kinds of soils and to sheep of all ages. There are two sufficient reasons why the disease is difficult to prevent: First, because the insect or fly that causes the disease eats but little or nothing during its life; and, second, because it deposits in the nostril of the sheep a living grub or larva. The disease is difficult to cure on account of its location. Turpentine and tobacco-liquid are sometimes administered with a syringe or by pouring in at the nose, but with not very good effect. The fly attacks the sheep more generally from the middle of June to September. Great injury is done to large numbers of sheep annually that are not destroyed. It is difficult to determine the per cent. that die, but the actual fatality is not greater than in some other diseases.

Perhaps more deaths occur from tape-worm than from any other disease, especially during wet seasons, when grass is abundant. It sometimes affects lambs at three months old, but does more damage to tegs just after weaning and previous to the appearance of good grass in the spring. Those affected become weak, pale, and do not grow; eat reasonably well, but irregularly ; drink abundantly and frequently ; in the first stages of the disease seem to lack power more than flesh. It has less effect on grown sheep. Those affected would appear to become wilted or shrunken ; the skin becomes very pale and thin; the wool does not separate from the skin as in other diseases. In the last stages the animal lacks blood. Occasionally they die within two or three months, but more generally they live for a longer period. I have doctored for this disease with very good results, having cured nearly all cases that were thoroughly tested. I use pumpkin-seeds, and administer by either feeding in other feed or by making tea. I also feed pumpkins, including the seeds. Information as to the cause of the tape-worm, and a preventive or cure, is greatly needed by the sheep-growers of the country.

We have no knowledge of the cause of the lung-worm—a name given for the want of a better perhaps. It affects young sheep in a greater degree and to a greater extent than matured animals. The worm is a small white one, and is found in considerable numbers in the lungs, or in the tubes connecting the wind-pipe with the lungs. The disease is less frequent than either of those named above, but the fatality is greater in comparison with the number affected. The symptoms are weakness, failure to eat, loss of flesh, and a cough. This disease is but little understood by the wool-grower.

Stricana or strichnia is perhaps a very incorrect name for the disease I wish to describe. It is caused by a very small worm, so minute, indeed, that it cannot be seen without the aid of a magnifying glass. It is believed to cause the sheep to pick or bite the wool from its sides, flank, and other parts, until the fleece becomes more or less ragged and wasted. The skin becomes rough and shows symptoms of disease. It is not contagious, but attacks sheep of all ages. It is more damaging in flocks that have been closely bred "in and in" for many years; indeed, this is the case with most diseases. As both a preventive and cure, wood and cob ashes with salt are used with partial success. We have seen sheep in Vermont and Massachusetts badly affected with this disease as well as in our own State.

A disease prevails in some parts of Ohio and Pennsylvania, and probably in some other States, that destroys large numbers of lambs annually. They are sometimes attacked by it at the age of three weeks, but oftener after they are two months old. The stomach, liver, and gall seem to be the only parts affected. There have been but few cases in this county, and we have no name for the disease. It is supposed by some to be caused by eating a poisonous weed, and by others by overfeeding on grass when too young. Wool is sometimes found in the lamb's stomach. The best and fattest lambs are frequently destroyed by the disease, with but little duration of illness.

Heavy losses are also annually sustained by diarrhea and dysentery. Proper food and management have more to do in preventing and curing these diseases than most others. The treatment and medicine that have been most successful are the same as those used in the human family for like diseases. A statement giving the best remedies and treatment of all these diseases would be received by thousands of sheep-raisers with great profit and many thanks.

Mr. A. G. GARDNER, Rutland, Meigs County, Ohio, says:

All farm-animals in this locality are comparatively healthy and free from epidemic or prevailing diseases. With fowls, however, the case is quite different. The losses have been heavy, and complaints are heard from every neighborhood of the terrible ravage of what is termed chicken-cholera. Whole henneries have been depopulated. No form of treatment appears to check the progress of the disease. I have never lost a fowl myself, and yet I raise from seventy to one hundred annually. I give my fowls full farm range, change my cocks each year by getting eggs from the best possible breeds, and select the best formed stock from these. They have high out-door roosting-places most of the season, but in cold, winter weather I confine them in warm, clean, well-ventilated roosts.

Mr. N. B. PETTS, Lincoln, Benton County, Missouri, says:

There is no disease existing among any class of farm-animals except among hogs, and among this class of stock there are several diseases, viz., measles, lung-fever, cholera, and worms. In this vicinity and throughout this county measles has prevailed to an alarming extent, and probably more hogs have died from it than from all other diseases combined. But nearly every farmer designates the disease as cholera. In measles the hog refuses to eat, lies much of the time in his bed, goes often to water to drink, but not to wallow, and grows poor very fast. The hog has a slight hacking cough. If the eruptions break out thickly all over the body the animal generally gets well; but if they do not, or after breaking out they should go back, the hog dies. There is a very offensive stench about their sleeping-places. Everything kept in a drug-store, and in quantities to astonish and alarm an alopathic physician, has been given, and the wonder is that so many have lived. All kinds of food have been given, but with no apparent beneficial results. The only thing claimed to have done any good whatever was a tea made from peach-tree leaves, limbs, or bark. This brings the measles out thick, and if the hog has a dry bed and is kept from water the chances are in his favor. The bowels should be kept open, and not more than five or six should be confined in the same pen.

For worms we give a teaspoonful of turpentine once a day for a week. From one-half to two-thirds of all hogs affected with diseases have died. I doubt if any have died of cholera.

Mr. E. D. RUSHING, Rock Point, Independence County, Arkansas, says:

A disease commonly called hog-cholera has prevailed extensively throughout this section. The symptoms are vomiting and purging, and death in a very short time. About two-thirds of those attacked die. Those that recover dwindle away and become almost worthless. Various remedies have been tried, but whether with any beneficial results is not known. A neighbor of mine, Mr. William H. Dood, after losing about one-half his herd, used tar water, which seemed to check the disease; at least he lost no more. The disease made its appearance in my herd in last September. It proved more fatal among my pigs and shoats, though I lost some bacon-hogs. I fed sulphur and copperas in swill, and in about four weeks the disease abated. I lost about one-fifth of all my hogs. The disease is still raging in some localities.

Dry murrain prevails to some extent among cattle. If given in time, the following remedy is said to be successful: Two dozen eggs well beaten and mixed with about an equal quantity of soap-suds. Milch-cows seem more liable to the disease than other classes of cattle.

Mr. L. N. HALBERT, Bonham, Fannin County, Texas, says:

We are not often troubled with diseases of a prevalent character among farm-animals, yet now and then we have glanders in horses, a disease which the old Texans denominate "Mexican distemper." When introduced upon a farm or on a range it proves very disastrous, and is beyond cure. I had it in my stock in 1871, and before becoming satisfied as to what it was, I used every remedy within my knowledge or that I could hear of. Finally I resorted to what I now believe to be the only remedy—powder and lead. It is much more fatal, does its work more rapidly, and is more to be dreaded among mules than among horses. I lost some of my best plow-teams of both mules and horses before I was able to arrest the disease by a change of stables, lots, &c., and the killing of those afflicted. I have also been greatly annoyed with fistula, a very bad tumor or rising on the withers. I have used many remedies, such as scarring with spirits of turpentine, lancing and putting in arsenic, burnt alum, concentrated lye, &c., but never succeeded in arresting but one case, and that at a very early stage. This I did by burning with a red-hot ring, circling the rising. A sure remedy is to rowel with a red-hot steel spindle through the cartilage of the neck, just missing the blade-bones. This operation never fails to cure.

I ought to have stated, while on the subject of glanders, that the symptoms are thickness of wind and stupor from three to five days before the nostrils begin to discharge. The discharges frequently begin in the right nostril several days before the left one is affected. The discharges are of a yellowish color, sticky if taken between the fingers, and becoming more abundant and offensive the further the disease progresses in its fatal work. The disease continues until suffocation ends the life of the animal.

Mr. JOHN M. CHAPMAN, Charleston, Mississippi County, Missouri, says:

The fistula, a terrible and offensive disease, makes its appearance on the withers of the horse just at the top of the shoulder-blade, at first upon one side only, but if the progress of the disease is not checked it will finally pass to the other side. A bruise of some kind is nearly always its cause. This the horse may receive in various ways, by striking the top of his shoulders in passing under a low stable-door, by bites from another animal, by rolling on stones or roots, or by an ill-fitting saddle. The disease is easily cured by the use of the following remedy: Take one-half bushel of may-apple root and pour over it about four gallons of water, and boil down to one gallon. Strain this, and mix with it about one-half gallon of old grease. Place the mixture on the fire and stew down to one gallon. During this process throw into it about one-half pint of salt, then let it cool, and it is ready for use. It should be applied with a mop or brush every morning, but the sore should be washed clean the night before. An application of this remedy will cure almost any case of fistula in from two to six weeks.

Founder prevails to a great extent in this locality. The first noticeable symptom is the restlessness of the horse and frequent shifting of the fore feet. The pulse is quick and his nostrils have a red appearance. The horse indicates his sufferings by heavy grunts. He does not stand long upon his feet, but cannot lie down in the usual manner. After making several efforts to do so he will rise up, turn round, change his position, and resume his feints to lie down. The remedy for this disease is to bleed freely without delay. Let the blood run free, and take at least a gallon of it. The object of this is to draw away the blood from the overloaded vessels of the feet. Always bleed in the neck. After this prepare a kettle of hot salt water, and drench with it as hot as the horse can bear it. Next bathe his feet and legs with it, and rub them well with a rough cloth. Make this application three or four times in the course of an hour, and then rub well around the edge of the hoof with turpentine. Do not attempt to work the animal until he gets entirely well. Another remedy is to pour the frog of the foot full of turpentine, hold it up, and burn the turpentine out. This is a little barbarous, but it is an infallible remedy.

Pole-evil is a tumor that comes on the head, or, more properly speaking, on the extreme forward part of the neck, just back of the ears. It is generally caused by being struck on the head by an enraged groom, and if it produces no other bad results it is sure to raise a large lump. I do not know of a case that ever caused death, but if not checked, the disease will render the horse unfit for use. The same treatment as in fistula will always effect a cure.

Mr. G. W. JOHNSON, Humboldt, Hunt County, Texas, says:

Blind-staggers in horses is perhaps the most fatal disease we have here. The remedy is to bleed freely from the neck, taking enough blood to cause the horse to show signs of faintness. Then give a drench composed of a tablespoonful each of spirits of turpentine, ammonia, and camphor, with about a pint of milk-warm water. Always drench through the mouth—never through the nose. Then burn tar, feathers, woolen rags, scraps of old leather, &c., under the nose. If this treatment is given nine cases out of ten will recover, if the horse is able to stand upon his feet when it is commenced.

Both dry and bloody murrain are very fatal to cattle in this vicinity. The best remedy for the first is a strong tea made of the common may-apple root, and for the latter saltpeter and guaiacum.

Cholera is the most fatal disease affecting hogs. The best remedy we know here is equal parts of guaiacum and copperas and Jerusalem oak seed, say a tablespoonful of each mixed in slops sufficient in quantity for five or six head of hogs. This has proved a good preventive as well as a cure.

Cholera is also fatal among all our domestic fowls. The best remedy I have tried is pulverized mustard-seeds. No particular quantity is prescribed, but it should be given freely. It will be found a cure as well as a preventive.

Mr. A. A. RUDY, Knob Lick, Saint Francois County, Missouri, says :

There are but very few, if any, diseases affecting horses, cattle, or sheep. Our main trouble seems to be with hogs and chickens. We have tried many remedies, some of which have proved of some value, and others of none whatever. It appears that every disease affecting either hogs or chickens is called cholera. Some of the swine are taken with a cough, and a swelling about the glands of the throat and neck, and generally live from one to ten days. Others have what I would call the measles. The skin becomes very red, and if they do not die, but on the contrary should recover, it will remain so for months. The following remedy, if administered in time, will be found an almost certain specific : Two and one-half pounds flowers of sulphur, one and one-half pounds pulverized copperas, one-half pound black antimony, and one pound of well-slaked lime. Mix well together, and then add one pound to a sufficient quantity of corn-meal or ship-stuff for twelve hogs. Put it in small piles on the ground, so that every hog will have a chance to get at it. As a remedy, it should be given every day until the hogs recover. After that, a like amount should be given once a month as a preventive. The hogs should, also, have all the wood-ashes they will eat. A good disinfectant may be found in lime. After slaking, take a broom, wet it in the lime-water, and sprinkle it over the beds of the hogs, until the ground is white, and about the coops and roosts of the chickens, if they are affected with cholera.

Mr. W. W. MURPHY, Madelia, Watonwan County, Minnesota, says :

With the exception of "blackleg," so-called, among cattle, I have never known of any epidemic disease among farm-animals in this vicinity. This disease carries off each spring, generally in March and April, a number of calves. There appears to be no remedy known for it here. I never knew of a case being cured. The loss in any one herd is not very large, but the annual average loss in the county is probably $500.

Mr. E. B. CASSILLY, South Charleston, Clark County, Ohio, says :

There is no disease prevailing among farm-animals in this locality except cholera among hogs. This disease has been very prevalent in this neighborhood and adjoining counties during the past summer, but has somewhat abated. Very few farmers have escaped its ravages. Probably one-half the last spring crop of pigs have died, and also a large proportion of the older hogs. I hear of one farmer who lost one hundred fat hogs. Not one in fifty recover. The symptoms are drooping of the head and ears and loss of appetite, heavy breathing followed by thumps, and purging and vomiting. The disease terminates fatally in a very few days. I have never known one to get well. Remedies without number have been tried, but without producing any good results. At least one thousand hogs have died in this township (Madison) during the past summer, and yet no remedy of any value has been discovered.

Mr. L. T. CURRENT, Brownsville, Saline County, Missouri, says :

Several diseases prevail among hogs here, but they are all called cholera. In some cases the symptoms are similar to those of lung-fever in the human family. Post-mortem examinations in some cases show the lungs to be destroyed, and in others gorged with blood. In other cases the hog is affected with vomiting and diarrhea. These symptoms indicate cholera, a disease which generally proves fatal, in many cases, within a few hours. It is my opinion, as well as the opinion of some of our best stock-raisers, that most diseases of hogs are caused by worms, for upon examination their intestines have been found to not only contain worms but to show holes in various places, which were evidently made by them.

The remedies used are as various as the opinions of the farmers concerning the cause of disease. Copperas is generally given, also sulphur, turpentine, and many other things. The best preventive so far found is black antimony and madder. It should be given about once a month. I use it and have never lost a hog. I also give my hogs coal and ashes, which also has a tendency to keep this class of farm-stock in health.

Fowls are also afflicted with a disease similar to cholera in hogs. As a remedy we use petroleum, onions, and common red-pepper. In the winter season these articles are mixed with thin feed, and in the summer in the water given them to drink.

Dr. P. A. FARIS, London, Laurel County, Kentucky, says:

Hogs are the only stock that we have much trouble with. They sometimes have dysentery, which I think is caused by eating clover, grass, and weeds, without a due proportion of grain, greasy slops, and salt to make digestion perfect. During the winter many are lost with mange, asthma, &c., which is caused by their sleeping in old straw and manure heaps. A few die from pleuro-pneumonia. But those who provide good dry leaf-beds for their stock, and feed them different varities of food, lose none.

Mr. SIMON DOYLE, Rushville, Schuyler County, Illinois, says:

In 1876 I lost 87 hogs by fever. They were invariably taken with a chill, followed by stupor and fever. There were no signs of cholera in any single instance, or any cough. Usually from four to six days, and sometimes from ten to twelve, intervene between the time of attack and death. I used many remedies, none of which were effectual in either curing or checking the disease. A large and strong sow, and the last one attacked, was the only animal that recovered. Some of my neighbors had hogs similarly affected. Others differed widely in the main symptoms, which were coughing, and bleeding at the nose, and death in from four to ten hours. In some cases worms were supposed to be the cause of death.

Mr. AMOS RILEY, New Madrid, New Madrid County, Missouri, says:

Hog-cholera, or "heaves," as some call it, is the most fatal of all diseases among farm-stock in this county. It is more fatal among the younger than the older hogs. Very few, if any, of those attacked recover. The symptoms are wheezing and cough, something like the thumps in horses. The duration of the disease widely varies. I have sometimes used corn, soaked in a solution of arsenic, with good effect. It is dangerous, however, to give this to pigs and sows. If the disease once gets into a herd it rarely stops until it cleans out (destroys) all the young hogs.

Mr. JOSEPH BORDERS, Painstville, Johnson County, Kentucky, says:

Farm stock in this locality is seldom affected with disease of any kind. Sometimes, however, that dreadful and very common disease known as cholera gets among our hogs and fowls and proves very destructive. Our hogs have escaped this year, but we have not been so fortunate with our fowls. The disease is general, and prevails at all seasons of the year. If there is any cure for it we have never been able to find the remedy. As to its cause I am ignorant. The disease scarcely ever reaches those fowls that are allowed to run in the woods and have a wide range.

Mr. SIDNEY GREIG, Vermillionville, La Fayette Parish, Louisiana, says:

Until within the last few years no fatal epidemic was ever known to exist among our domestic animals. But now, on the return of the spring and summer months, we have a disease which attacks horses and mules, and sometimes cattle and sheep, and is very fatal. From the rapidity of its action there is rarely time to administer any remedy, and if any is given, not knowing the nature of the disease, it is only a lick in the dark—death is certain. The disease is endemic in its nature, confining itself one season to a certain locality, when it will disappear, and the next season it will make its appearance several miles off. I have been a careful observer of this disease, for I have been one of the sufferers from it, and will give you as exact a diagnosis as I possibly can. The symptoms are drowsiness, loss of appetite, and fever. As the disease advances the animal becomes restless, and walks continually, although without seeming to suffer any great pain until the last hour preceding death, when the agony is intense and pitiable to behold. In the last stages a profuse sweating ensues, and the animal shakes as if in a congestive chill, and soon fall and dies. A *post-mortem* examination reveals the whole internal organs a mass of congestion, and the heart, liver, lungs, and intestines covered with a yellow, jelly-like substance. Neither a preventive nor a cure has as yet been found for the disease. The only preventive seems to be

found in the removal of the stock until cold weather sets in. After careful consideration I am fully convinced that it is a malarial disease, similar to that which affects the
human family, but of a much more violent character. I have no doubt if like remedies could be applied in the beginning of the attack many animals could be saved.
The causes, in my opinion, have the same origin as in cases of malarial or intermittent
fever which afflicts the inhabitants of Lower Louisiana, viz: The want of proper
drainage, the use of impure drinking-water, and the lack of proper care, especially of
our work animals, for it is this class that suffer the most. The duration of the attack
is from six to twelve hours.

When this disease makes its appearance, had we a competent veterinary surgeon to
make a careful investigation of its symptoms from the first stages until the final act,
and a scientific *post mortem* examination held, there can be no doubt but it could be
robbed of its present terrors, and many a poor man's heart caused to rejoice thereof.

Mr. HENRY M. DARNALL, jr., Gayoso, Pemiscot County, Missouri, says:

We have but little disease among farm-stock here, except blind-staggers in horses
and a disease called cholera in hogs. Several horses have died of the former during
the past few years, and the cholera has at times been quite fatal to the hogs. The
first symptom of the last-named disease is a slight cough. Their eyes soon become inflamed, and they appear to get sore all over. They often have a number of abscesses
and tumors on them. The disease is of several weeks' duration, and is generally very
fatal. As a remedy I have used with good success one tablespoonful of carbolic acid
in slop sufficient for twenty head of swine, giving it to them once a week.

Mr. A. M. ELLISON, Beaver, Douglas County, Missouri, says:

There are no diseases of any kind existing among farm animals in this locality except
cholera in hogs. The first symptom of the disease observed is a refusal of the animal
to eat. In some cases the teeth seem to be sore, so much so indeed as to prevent the
animal from chewing corn. They often linger from ten days to two weeks, but the
disease generally proves fatal within that time. Those that do recover generally shed
most of their hair. A few animals are affected with a cough. We have no remedies
for the disease.

Mr. M. S. BARTRAM, Ironton, Lawrence County, Ohio, says:

The disease among the hogs of this county is generally known as cholera. It has
been very prevalent this year, and the losses have been quite heavy. One farmer lost
150 head, including stock-hogs, which was about all he owned. The disease does not
seem to be so prevalent among fat hogs nor so fatal as it is to those in moderate condition. Those running at large seem most liable to the disease. No remedies are used.

A disease among chickens is very prevalent, and is not confined to any particular
section or locality. It is supposed to be cholera. The only remedy so far used has
been black-pepper, but without beneficial results.

Mr. C. J. C. BOYNTON, Pulaski, Williams County, Ohio, says:

For a number of years past we have been troubled with a disease known here as
chicken-cholera. Three years ago this fall I had 150 head of pure breed and half-breed
Light Brahmas. They were attacked with this disease, and in about a month or six
weeks I lost over 100 head. When first attacked the head of the fowl would turn purple, and it would begin to droop and mope about. In a little while diarrhea would set
in, and the excrements would be of a greenish color. The fowls lived but a short time
after the first symptoms showed themselves; some would die in a very few hours,
while others would linger for a day or so. Since that time the disease has visited the
flocks of about all my neighbors and with about as fatal results. We have found no
sure remedy for the disease. We tried indigo in the water they drank, a solution of
white-oak bark, and many other things, but without apparent benefit.

Mr. PETER HOLLOWAY, Monclova, Lucas County, Ohio, says:

A very fatal disease has prevailed among hogs in this vicinity. Mr. H. L. Holloway,
of Springfield, had 90 head attacked with it. It seemed to partake of the nature
of a lung disease, as it was attended with coughing and a high fever. The teeth also
appeared to be tender and sore, as the animals could not bite corn off the cob. Those
that died were almost completely covered with sores. The cause or origin of the disease is unknown. One theory is that they contracted the disease by lying and wallowing in the mud and water from the overflow of an artesian well strongly impregnated

with sulphur. The disease was first observed September 1, its greatest fatality occurred October 10, and the last death on December 20. Of the 90 head attacked, two were shot, three recovered, and all the rest died. The skin on those that recovered nearly all peeled off. They were in good condition up to the time of attack, having run in blue-grass and clover pasture during July, August, and September. Those that were afflicted were carefully cared for. The remedies used were arsenic, calomel, charcoal, sulphur, copperas, fresh meat, and carbolic acid, but without any beneficial effect. The age of these hogs ranged from four weeks to five years.

The disease is regarded as contagious, for the following reason: About the time of the commencement of the disease, but before he was aware of its existence, Mr. Holloway sold a sow and five pigs to a Mr. Graham. They were taken to a distant neighborhood and put into a pen with another pig. Soon after they were taken sick and died, as also did the pig which was confined with them. In an adjoining pen were six fattening hogs. One of these was taken sick, and in order to prevent the further spread of the disease Mr. Graham killed the balance.

There has been a very fatal disease prevailing among the chickens in this neighborhood, which is variously called the roup, the hen fever, and the hen cholera. Fowls attacked with it appear stupid at first; their combs turn purple, and they gape frequently. They have been known to die within four hours after the first symptoms were observed, and seldom live more than a day or two. Guinea-fowls seem to suffer from the same disease. The most successful remedy used was a strong decoction of white-oak bark, made by boiling in water and mixing with cornmeal, adding about two-thirds of a teaspoonful of cayenne pepper to the quart of feed. They also placed the ooze in vessels where the fowls had easy access to it. This seemed to check the disease at once.

Mr. C. LEWIS, New Vienna, Highland County, Ohio, says:

The hog is by nature a very healthy animal, and should be the same in his artificial or domestic state. Therefore, in investigating his present condition, reference should be had to his original habits and surroundings; and the nearer we can approach this in his domestic condition the better. We find that in his natural state his home is in the forest, where he can roam at will and indulge his appetite in partaking of its productions in the form of roots, grasses, herbs, fruits, berries, nuts, &c., in their proper season and natural purity; making his bed in leaves by the side of logs or other temporary shelter, changing the same at pleasure, and reconstructing his bed out of new material, and all the time using his "snoutish" proclivities to the full bent of his instinct. Thus we find him a healthy, and in his maturity a powerful animal. Now, the nearer we can conform to these first principles or habits of the animal the better, for the preservation of health and prevention of disease is far better than all the remedies known or unknown. In his natural condition we find him comparatively free from all filth, dust, and foul air, making his bed out of leaves or grass on the ground, sleeping few in a bed, and drinking pure water. And now, as to his domestic condition, I will not say habits, for he is no longer free to exercise these; and right here is the first line of demarkation between health and disease, and must be so considered if we wish to arrive at the truth of the matter. The cause of the disease seems to be more easy to point out than to remove. In the first place, there are, as a general thing, too many hogs kept together in the same inclosure, which gives them an opportunity to "pile up" in their beds when the weather is cold and stormy, becoming not only overcrowded but over-heated; thus laying the foundation for disease by disturbing their normal condition. By this confinement they are also compelled to a greater or less extent to be ever present with their waste matter, which at certain seasons is more detrimental than at others; hence at such times they are more liable to attack by the so-called epidemic diseases.

Another cause of derangement and disease is dust, which is generally most abundant at the season when the waste matter is most offensive and detrimental, thus producing a double aggravation of the cause of disease. Another productive cause is the habit of keeping the same stock of hogs on the farm for a number of years, even when there is an annual change of male hogs. If a change of pasture will make fat calves, an entire change of stock will certainly produce better and more healthy hogs, other things being equal.

Now as to the diseases to which the hog is subject: Though naturally healthy they can secrete a mountain of disease, and it does seem that a diseased hog is the worst diseased animal on the face of the earth. There appears to be an epidemic disease of the lungs, commencing with a cough and followed by loss of appetite, general debility, and finally running into something similar to lung-fever, which is generally fatal. The principal producing and exciting causes of this disease appear to be dust, too many occupying the same bed, foul air and exposure to cold, wet storms. (The disease seems more common among pigs and shoats.) There is also a disease of the bowels, which might be termed cholera or diarrhea, and seems to prevail more extensively among hogs fed on dry corn. I have never known a hog fed on soft or cooked corn to be afflicted with the disease. There is still another disease, that of the spine or hind legs, which appears to differ from the so-called "kidney-worm," and is not unlike rheumatism as it affects the human family. This is generally fatal. There are also diseases of the liver, intestinal worms, &c.

Mr. J. M. ANABLE, Naples, Ontario County, New York, says:

We have been very much annoyed by abortion in cows. It seems liable to come on at any time. No cause has been discovered, and of all the remedies that have been tried none have proved of the least benefit. When it gets into a herd it generally affects from one-third to one-half.

There have been a few cases of blackleg among calves that were in good condition. About all of the cases proved fatal. No preventives or remedies have been found.

Garget or udder-ill has been the source of much annoyance with our best cows. The disease affects the udder and causes the milk to become lumpy; if the disease is severe it becomes bloody, the teats swell, and hard bunches appear on the udder. As a remedy one quart of warm lard and one-half pint of molasses given as a physic, together with frequent bathings of the bag with cold water and drawing off the milk three or four times a day, will be found beneficial. If inflammation should be great apply fomentation to soften the udder, and use a mild liniment or ointment. About 20 per cent. of our cows are affected by this disease, and about 10 per cent. of these are rendered unfit for dairy purposes.

Mr. H. H. WILSON, Salem, Livingston County, Kentucky, says:

In 1874 I had ten head of shoats that took the cholera, and eight of them died. I tried many remedies, among others tea made from May-apple root, red pepper, asafetida, &c., in slops. I also gave them soft-soap and salt mixed with wheat-bran, at the rate of about one gallon of soft-soap to sixteen head of hogs. Only those hogs that were able to eat the preparation recovered. I have since given soap and red-pepper tea as preventives, with, I think, good results. A neighbor of mine, Mr. Phil. Graham, who is one of the most successful hog-raisers in this county, says that pokeroot tea will cure cholera in hogs.

Mr. R. L. RAGLAND, Hyco, Halifax County, Virginia, says:

Diseases among hogs in 1877 were unusually prevalent. More than half of those attacked by cholera died. Measles and quinsy were not so fatal. Measles was th most prevalent disease during the past year. Cattle are annually subject to distemper, a violent grade of fever that prevails more or less every year. More than half of the animals attacked by the disease die. We have no reliable remedy for it, but have found a preventive that has proved very efficacious. It is this: To a bushel of red clay add one gallon of salt, four ounces of saltpeter, and two ounces of sulphur. Mix, adding sufficient water to make the mass of the consistence of mortar, and put it in troughs for the cattle to lick.

Mr. J. K. KIDD, Kiddrige, Osage County, Missouri, says:

Hog-cholera, so called, has been and still is quite prevalent in this section of the county. On the first indications of the disease the hog sometimes coughs, but not always. Sometimes they are constipated and again quite lax. They refuse food, go about in a kind of listless, drooping manner, and apparently have fever. Several have died on my place. They were not confined in pens, had an extensive range, selected their own beds, and in doing so avoided the hog-house. A variety of remedies were given them. Sweet milk and allspice, poke-root juice administered in slops, coal-ashes, sulphur, &c., were given, but with little apparent benefit. A majority of those attacked die.

Chickens are also subject to a disease called cholera, for which no specific has been discovered. Those affected seem stupid and drooping, the crop and liver swell, and they die suddenly and by dozens.

Mr. AMOS TODHUNTER, New Martinsburg, Fayette County, Ohio, says:

The most prevalent disease in this locality is among hogs, and is called cholera. As it has not visited my farm, I asked the assistance of Dr. M. Todhunter, who is familiar with the disease, and he responds as follows:

The first symptom is that of fever of a typhoid form. Then follows a disturbance of the head, lungs, and bowels. When the lesion was on the brain sores would appear about the head, and the ears would ulcerate and emit a very offensive stench. When seated on the lungs there was an almost constant cough. When dead the lungs of some were found to be almost rotten, and smelled so bad that it was difficult to handle the carcass. In the absence of the above symptoms the animals seemed to live longest; that is, longer than when the lesion was on the bowels. The bowels ulcerate, and the ulcerated matter passes off with the fecal discharge. Constipation prevails in all cases. Those that are relieved earliest of this difficulty are the most apt to recover.

I tried all the remedies known, and they were very numerous, without much apparent good. The best treatment I found was to change frequently the locality of feeding, and to give them a good supply of salt and ashes, mixed with bran. This I fed whether the hogs were sick or well. I put the sick ones to themselves on a grass-lot, and fed lightly with slops, putting sufficient sulphate of magnesia into the slop to produce an operation on the bowels. I continued feeding lightly until there were signs of returning appetite, when I commenced gradually with corn.

I am of the opinion that over-feeding in the start is the cause of these diseases in swine.

As to my own experience I will say that I raise from 50 to 100 head of hogs annually on my farm. It has been my practice to change their locality quite often during the course of the season, and to give them all the slops and soap-suds from the kitchen and wash-house. I also give them ashes and cinders from both coal and wood, adding salt, and occasionally a little sulphur, which I think has a tendency to destroy the lice which infest them during dry weather. I do not house them unless the weather is very inclement. They seem to thrive best when they have plenty of leaves to bed in. Next to this is corn-fodder, wheat and oat straw not being so good. With this treatment my hogs have remained healthy, while those of my neighbors have been attacked and died of the various diseases to which they are incident.

Cholera also prevails in this locality among domestic fowls. Some farmers have lost very heavily. Many remedies have been used, but without apparent benefit.

Mr. W. A. HELM, Sugar Grove, Butler County, Kentucky, says:

The principal disease to which horses are subject here is a contagious distemper, which is most prevalent in the spring of the year, but frequently returns in the fall. The disease prevails throughout this State, and perhaps others. The first symptom is a slight cough, which continues until it renders the animal unfit for use. Loss of flesh, stupidity, and apparent laziness are characteristic. If the animal does well, after coughing for some days, it will eject large lumps of matter from the nostrils; but if the disease assumes a fatal form the throat becomes swollen, until breathing is almost stopped. It is not often fatal, but it frequently affects the breathing of the animal to such a degree as to injure its sale and use.

The prevailing disease among hogs in this section is what we call cholera. Whether it is the real cholera or not I do not know. The first symptom is a refusal of food. The lungs, lights, and liver all seem to be affected, and breathing is rendered very difficult. The disease has been very fatal in many of the Middle States, as it has here. The animals rarely, if ever, recover. When a cure does seem to be effected the hair always remains rough and of an unhealthy color.

Mr. GEORGE H. JUDSON, San Antonio, Bexar County, Texas, says:

The facility with which horses and cattle are raised here, without any care other than marking and branding, has bred a carelessness among farmers and stock-raisers that is truly deplorable. Trusting to nature entirely to provide food for their stock, when a cold winter follows a droughty summer, thousands of cattle and many horses die of starvation. The introduction of railroads has brought a new class among us, and they are bringing a better grade of cattle with them. Lands are being fenced and stocked, and some care is beginning to be observed in the treatment of farm-animals. Whether disease will follow is yet to be determined.

I have been a raiser of sheep for several years past. The only disease seriously affecting them here is apoplexy. Our oldest and fattest animals are generally the ones to suffer. From a small flock of 540 I this year lost 110 head, nearly all of which were wethers and excessively fat. There are no previous symptoms. To an inexperienced shepherd the sheep appears remarkably well, and apparently very happy, often frisky, when he suddenly makes a leap into the air, falls, and in less than three minutes' time is dead. This disease only occurs in excessively hot weather when water gets low, or when they have to be driven some distance to water. I have heard of no remedy of any value. Some starve their sheep by keeping them in their pens until eight or nine o'clock in the morning, and then folding them early in the evening. This may do, but I doubt it.

Last fall we had a new disease among chickens. Something like a pimple or wart appeared on the heads of the young chicks, and after a few days the chick would lose its sight, and then wander aimlessly around until it starved to death. These warts made their appearance on the eyelashes and about the bills. Copperas-water was freely used, and all the adults saved, but the young chicks were not benefited. In fact, they were not much cared for, as they were a cross between the common fowl and Brahmas. Had it occurred among the full-bloods, in all probability they would have been saved. In all other respects the chicks were in good health, as they had an excellent appetite.

Mr. GEORGE W. MINIER, Minier, Tazewell County, Illinois, says:

Our chief trouble by way of disease is with swine. The disease is known as cholera, but doubtless the cause of it is an intestinal worm or parasite. Sometimes the lungs are affected. From the first the animal refuses food, it is mopish, coughs, and sometimes has what is vulgarly called "thumps," *i. e.*, shortness of breath with quick beating sides. One of our best remedies is indigo dissolved in water. Our domestic fowls are affected much in the same way, and people give the disease the same name. Our more hardy and early varieties, such as Dominique, Game, &c., are seldom sick, and it may be that our finer varieties brought with them the germs of the disease.

Mr. R. H. LEE, Duvall's Bluff, Prairie County, Arkansas, says:

The principal diseases affecting cattle in this county are known as dry and bloody murrain. Dry murrain, which is supposed to be caused by insufficient supply of water, is cured by large doses of calomel. There are many other remedies for it. We have no successful remedy for bloody murrain, and very few animals attacked by it recover.

Horses are seldom affected with diseases, but last summer a neighbor lost five head with a malady previously unknown here. The animals were taken with a limping in the fore legs, but recovered from this in a few hours, when a high fever set in. The horses did not lose their appetites, but took feed liberally. Every case, and there were a good many in the neighborhood, proved fatal within from three to five days. Many different prescriptions were given, but they all failed to give relief. The animals did not appear to be much distressed at any time. They generally died very suddenly and without a struggle.

Fowls have what is called chicken-cholera, a disease which is almost invariably fatal. The liver is generally found to be very much enlarged. I have tried calomel, quinine, rhubarb, cayenne pepper, copperas, sulphur, and indeed almost everything else, without success.

Mr. JAMES BOWLDEN, Will's Point, Van Zandt County, Texas, says:

Most horses, but particularly young stock two years old, are, in the winter and spring of the year, attacked with a disease similar to the epizootic, and many stock-raisers think it one and the same disease. It is generally known here, however, as the distemper. The symptoms are cough, swelling of the glands of the neck, and a profuse discharge from the nose of a thick, green-colored matter. It is sometimes fatal, but rarely so, and the animal often recovers without any help. All that seems necessary is good warm stables and careful feeding. Spanish fever attacks many animals brought in from other States. All imported animals are subject to this disease until they become thoroughly acclimated.

Cattle are subject to a disease called murrain, which generally proves fatal. Various remedies have been tried, but with little success. Imported stock (short-horns) are subject to a disease called by some Texas fever and by others Spanish fever. The disease is very fatal, as but few animals survive. No satisfactory treatment or remedy has been found.

Hog-cholera seems to be more fatal than any other disease affecting farm-stock. The symptoms are loss of appetite, blindness, dullness, and weakness in the loins. Kerosene-oil and turpentine have been used quite successfully as a remedy when administered during the first stages of the disease. Many suppose the cause of the disease is from worms in the kidneys, as these organs are found, after death, to be more or less affected. Chicken-cholera is also quite prevalent and fatal. We have no preventive or cure.

Mr. JAMES W. TERRELL, Quallatown, Jackson County, North Carolina, says:

Here in the mountains of Western North Carolina, by far the greater part of the income of the people is derived from the sale of horses and cattle, particularly the latter, while hogs, sheep, and poultry contribute in a smaller proportion. As we work our horses and mules while young, and sell a large proportion of them after maturity, it is only in rare instances that one ever dies. The epizootic swept along in the fall of 1872, but by the time it reached us it had assumed so mild a type as to do but little harm, and it has not since reappeared. Our young horses sometimes have something like influenza, but it seldom proves fatal, the animals recovering without treatment. What is known as "bots" or "grubs" is the only really formidable disease that attacks the horse here. The symptoms are restlessness, loss of appetite, the eyes appear weak and the whites enlarged, or more apparently visible, the gums and lips pale and clammy. The animal frequently turns his head toward his flank, lies down frequently, but does not roll as with colic. As a remedy I can scarcely think of anything in the whole veterinary practice that has not been recommended. Sage-tea followed by a purgative, sweet milk and molasses, spirits of turpentine, a bluestone pill, are among the most commonly-applied remedies. I look, however, upon a copious drench, say a quart, of a strong decoction of the common garden tansy as the most efficacious. I do not give it as a specific, but I have not yet known it to fail, if given in the early stages of the disease. As a preventive, keep a cloth saturated with hog's lard in the stable during the months of August and September, and occasionally or daily rub the horse lightly with it over the parts where the "bot-fly" deposits its eggs or nits on the hair. These nits by some means get into the horse's stomach, and hatching there produce the grub. Grease kills the egg and prevents its hatching.

Hogs have cholera, or a disease which we call cholera, that in the last two years has cut our hogs down below the demand for home consumption. The symptoms are loss of appetite, disinclination to move, vomiting, diarrhea, eruption of the skin, loss of hair, and. of course, great loss of flesh. It is very fatal, killing, I think, over half the animals it attacks. It seems to be epidemic. I do not think it is contagious. What causes it? A writer in Illinois—*vide* Country Gentleman—says an exclusive corn diet; but here it attacks equally our hogs in the wild mountain range with those raised on the farm, those fed on kitchen-swill, garden-vegetables, or by a mixture of all these things. It also attacks all breeds from the Berkshire down to our native razor-backs, and all the intermediate grades. We have no remedy. A good many things have been tried, and sometimes the animal gets well, but I believe as large a proportion without as with treatment. My own experience, corroborated by that of some of my neighbors, is that a plentiful supply of fresh wood-ashes and charcoal, with a little salt, kept where the hogs will have continual access to it, is a preventive. One would be surprised at the avidity with which they will eat this mixture. I lay great stress on this preventive, for I do not remember that I ever had an animal attacked with the disease when it had been supplied with the mixture, and, as a verification of the adage that "an ounce of prevention is worth a pound of cure," I have never had a hog to recover from the disease.

We also have chicken-cholera, but I know neither remedy nor preventive. I only know the chickens refuse to eat, droop a few days, and die. A neighbor tells me: "Feed your chickens on dough made of corn-meal and soft (lye) soap and they will not have the cholera." It is simple and worthy of trial.

Mr. ALBERT BADGER, Nevada, Vernon County, Missouri, says:

Last year this county lost many thousands of dollars in horses, cattle, and hogs, and this year seems to be no exception, as the same diseases have prevailed to a greater or less extent every year, for thirty years past. I believe this can be said of every county in the State. At least 80 head of horses have died in this county since October, 1877, from the effects of eating worm-eaten corn, and in all probability as many more will die before grass comes in the spring. It is true this loss can be avoided by carefully removing all worm-dust from the corn before feeding, but many never know the danger until too late. Others, boys and hired help, although often warned to be careful, are just the opposite. The symptoms of the disease are various; sometimes it results in blind-staggers, crazy fits, stupidity, and general prostration; sometimes they will sit for a long time like a dog. I believe, from the start, they are partially blind or entirely so. The disease has never been cured, and we sorely need an antidote for this worm-poison.

We also lose quite a large per cent. of horses every year by bots. The fly which produces this grub is very plentiful in prairie countries. Specifics are used which sometimes succeed in causing the worms to let go, otherwise the horse dies. I lost one of the finest animals in this part of the county during the past summer, within fifty minutes after he was attacked. One of my neighbors lost two last week, and so they go.

The most troublesome disease among cattle, which yearly hangs to us, is blackleg, for which we have no preventive or cure. The disease is most prevalent and fatal among calves and young stock. It invariably attacks and kills the fattest and most promising calves, and leaves the poor and runty ones. Either a preventive or cure would save millions of dollars annually to the people of this State. I might as well state here that a drove of Texas cattle slipped through this county in September last, and left a disease which killed at least $2,000 worth of native stock. I lost five head of cattle myself by it, and I can say with all truth that we would all feel much safer if we had a remedy for this terrible scourge.

Your department should never rest until Congress furnishes the means to sift the terrible disease of hog-cholera to the bottom, and through science and experiment find either a preventive or cure. There are more hogs that die of this disease every year than are consumed by the people of the Western States. Our farmers could afford to pay one-fourth of the national debt to be relieved of this one disease; and if they had certain cures for poison by worm-dust, for bots and blackleg, the amount saved in twenty years would pay another fourth.

Cholera is very destructive to all kinds of domestic fowls. I have recently lost over one hundred chickens by it, and one of my neighbors as many turkeys. The loss was equal to 90 per cent. of our flocks. We have no remedy for the disease.

Mr. W. L. ROBBINS, Mayfield, Graves County, Kentucky, says:

For the last twelve months we have been suffering from a disease called hog-cholera. Examination after death reveals an affection of the lungs and intestines. The hogs live but a short time after they are taken with the disease, and generally die in their beds and apparently without much suffering. We have been unable to find a remedy. Copperas, arsenic, sulphur, salt, and wood-ashes are used as preventives, and it is thought with beneficial results. Not over 10 per cent. of those attacked recover.

We also suffer to some extent with chicken-cholera. Alum administered in wheat-dough is regarded as both a preventive and cure, but it cannot always be relied upon as either.

Mr. GEORGE HUNTER, Carlinville, Macoupin County, Illinois, says:

Presuming that breeders of the several classes of farm-animals and fowls will respond to your circular-letter with such information as concerns mainly the class with which severally they are most conversant, I shall confine myself to a few pertinent facts coming under my observation as a breeder of swine. I state upon careful inquiry and personal observation in my own neighborhood and adjacent localities, that about 20 per cent. of the entire hog crop, in numbers, die annually of the various diseases incident to swine. Of this loss about 15 per cent. is probably due to hog-cholera, and the remaining 5 per cent. to other (practically) obscure ailments. In this section of Illinois, which is one of the heaviest corn and pork producing regions of the West, I should estimate the loss annually, in dollars, by the diseases among swine, as equal to about one-fourth of the entire hog product. From the mass of general statistical information to which one properly turns in this connection, it may be inferred with a reasonable certainty that in this class of animals alone the country at large sustains an annual loss of at least $15,000,000 by the ravages of disease, the State of Illinois bearing perhaps $2,500,000 of the loss as her share.

As to measures of prevention or treatment (inquired of), whatever may be known to veterinary science, or possibly professional skill, nothing, by way of general relief, has been accomplished. No precautions of a general character, to prevent the spread of contagion; no concert of action for the purpose of disinfection, has ever, so far as I know, been attempted. And basing my observation upon the magnitude of the interest involved, the wide-spread character of the evil, and the highly contagious and fatal character of the disease prevailing, I respectfully submit that no amount of private enterprise or personal effort can avail for the protection of the public good, and that no system of prevention or disinfection can ever be adopted, of a sufficiently general or uniform character, to be effective in protecting the public interests in this matter, unless that system rests upon the authority of government, and an adequate fund, such as Congress alone can provide.

It can scarcely be of service to increase the enormous mass of confused, illogical, and contradictory reports of diseases and treatment which are found at every hand, as enough already appears in these accounts to show that nothing more is to be hoped for in that direction. Facts enough have been laid before the public, observations and conclusions enough, bearing the test of scientific experiment, have been made, upon which to predicate the belief that a competent commission, having the requisite authority and funds, could easily frame and establish a system of simple sanitary measures, which, being generally applied to this class of farm-animals alone, would result in a vast saving to the country, even though no specific cure for that dreadful scourge, "hog-cholera," should be discovered. Let the appropriation be made, let the commission be authorized, and let its investigations be thorough and searching. This I take to be the general view of the subject on the part of those who have given the matter attention.

Mr. M. BLEVINS, Maysville, Benton County, Arkansas, says:

In reply to your circular-letter of recent date, I would say that out of 110 head of cattle I have lost 23 with a disease we call murrain, and out of 70 head of hogs I have lost 20 with cholera, so called. I think the average loss among the farmers in this county is about the same. We have no remedy for either disease.

Mr. M. A. KNIGHT, Middleburg, Clay County, Florida, says:

For many years past a disease called staggers has prevailed among horses in this locality. It is a disease of the brain, and in my opinion is brought on by over-work, or in permitting the animal to graze during the heat of the day. The symptoms are an entire loss of appetite, costiveness, restlessness, a disposition to walk and seemingly not caring where, and oftentimes describing a circle. As a remedy bleeding freely in the hind parts is practiced with considerable success. I prefer to cut off the end of the tail, and if necessary cut off a second time if the first operation does not give a free and continual flow of blood. Then bathe or rather pour cold water on the head until the disease is arrested. This should be followed by a good dose of Epsom salts, say one-fourth of a pound dissolved in water, and repeat if a free movement of the bowels does not follow the first dose.

In the early recollection of the writer, say twenty-five or thirty years ago, this disease was very fatal to horses, probably not more than one in twenty being saved by treatment then in vogue. Since the foregoing remedy has been practiced from 50 to 75 per cent. of those attacked by the disease recover. The disease is prevalent only during hot weather, and seems to principally affect the brain. It is doubtless brought on by exposure to the sun either while working or grazing.

Mr. WALTER BARNES, Larissa, Cherokee County, Texas, says:

Among hogs the principal diseases are known as cholera, quinsy, and kidney-worms. With cholera the symptoms are a constant retching, with slight mucous discharge, and staggering and apparent blindness. Death generally ensues within from three to twelve hours. The disease is very fatal, and but few of those attacked recover. I assisted a short time ago in opening a hog that had cholera last fall (1876), and during this year (1877) which had occasional spells of loss of appetite, without any other apparent ailment. The body, entrails, pleuro, and vitals of the animal were all grown together, and had to be separated with the knife. The liver was twice its natural size.

With the quinsy there is a difficulty in breathing and swallowing, which continues until the animal dies. I know of no remedy for either cholera or quinsy.

With kidney-worm the animal shows weakness in the hind legs, staggers, and unless relieved gets down in the loins and drags its hind legs on the ground until it dies. As a remedy give small doses of strichnine twice a day for three days, and pour a teaspoonful of spirits of turpentine on the loins twice a day.

Among fowls the cholera has been prevalent in many localities. Those dying of this disease are found to have an enlarged liver. Sometimes this organ is increased to two or three times its natural size. I believe fowls need salt as much as farm-animals, and mine get it. To my knowledge we have never lost one by disease, while all my neighbors lose more or less.

Dr. ANDREW J. WILLIS, Saratoga Springs, New York, says:

The only disease that has prevailed here among farm-animals during the past year was intestinal fever in swine (cholera). All cases proved fatal. The average duration of the disease was two days. No treatment seems to have been given. I saw none of the cases, and the only information I have I received from eye-witnesses. The symptoms and lesions described were those of hog-cholera. There were seventy cases in all. From the information I have been able to glean I am of the opinion that the disease was not of a contagious character; but I think unwholesome food contributed largely to its diffusion, if not to its development. The hogs were fed with food from the large hotels in this place, which usually contains a large per cent. of green vegetables, which, in warm weather, rapidly undergoes decomposition. I am informed that the feeding-troughs were never cleaned out, though swarming with maggots, and that the pens emitted a terrible stench. From this it will be seen that the swine were not kept in the best of hygienic conditions. The outbreak cannot be traced to contagious influences, nor can we say it appeared spontaneously, although we must concede that some cases probably originated spontaneously. No doubt the unwholesome food favored the development of the disease by loading the blood with deleterious organic matter, and so brought about a susceptible condition of the system.

Mr. J. L. SEARS, Valley Mills, Bosque County, Texas, says:

We lost a few of our horses and mules last winter by a disease called blind-staggers, and this winter a good many work animals, both horses and mules, have died from a similar disease. It is supposed to be caused or superinduced by worm-eaten corn. I lost one horse and had several others attacked by the disease, but relieved them by smoking with pine-tar, woolen rags, and red pepper, and by giving them large doses of bromide of potash. I also bled in the neck. A very strange thing about this disease is the fact that every horse attacked loses the sight of his left eye, yet you cannot detect any difference in the appearance of the eyes. They both look natural, yet the animal cannot see one particle with the left eye. They will not see you if you approach them from the blind side, but as soon as you show yourself on the right side they become alarmed, wheel from you, and throw themselves against the walls of the stable with such force as often to knock themselves down. When in the lot they will continue to turn round in a circle until they fall, and then, unless promptly treated, will die in a few hours. Out of twenty attacked in this neighborhood ten have died. Since quitting corn as a feed and substituting oats my animals have done well.

Swine have not done very well for the past two years. A great many have died from a disease called cholera, but I am of the opinion that a great many more die from the effects of eating cotton-seeds and cockle-burs than from cholera.

Mr. JOHN PITMAN, London, Laurel County, Kentucky, says:

The most troublesome disease we have to contend with is cholera among hogs. The losses were very heavy during last fall. With the cold weather the disease has disappeared, and no animals seem now to be affected with it. The first symptom of the disease is a stiffness of the limbs, the animal moving about like a foundered horse. The eyes become watery, the hog vomits frequently, and the excrements are bloody. The hog generally dies within twenty-four hours. The best treatment is to change their quarters frequently, and feed them turnips (tops and roots), potatoes, pumpkins, and such things. If they can be induced to eat, the chances are favorable. If they will not eat there is no need of giving them medicine. I had eight cases of cholera in the fattening-pen last fall. After five had died I turned the others out into a lot and fed on turnips, giving slops occasionally, in which I put a little copperas and salt. The three sick ones recovered on this diet. At the same time I lost about twenty pigs that were running at large. I gave nothing in the way of medicine, except calomel to one, and it died. The symptoms were stiffness, blindness, coughing, and watery eyes. The whole lot died within a period of twenty days.

Mr. W. B. FLIPPIN, Yellville, Marion County, Arkansas, says:

A few cases of hog-cholera are reported in the county, but whether the hogs die of cholera or from the effects of eating cotton-seeds where cattle are fed is hard to determine. I am sure that more die in this locality from the effects of feeding on these seeds than from other causes or with other diseases.

Professor JAMES LAW, of Cornell University, Ithaca, N. Y., says:

A life-long study of the diseases of domesticated animals has convinced me that government interference in such matters is altogether uncalled for, excepting in the case of such maladies as are communicable by contagion or otherwise from animal to animal, or from animal to man, and *vice versa*, and the existence of which in this country, or in one with which we have commercial relations, endangers our live-stock interests or the health of our people. Apart from these, the duty of the Executive will be sufficiently fulfilled in the foundation and maintenance of a fully equipped veterinary college and experimental station, similar to those of Continental Europe, and under such supervision and control as will protect it against those debasing courses which proved the ruin of the two earliest American veterinary colleges (Boston and Philadelphia). Such a school would be of unspeakable advantage in investigating the diseases indigenous to our different States and Territories, and in sending out men on whose knowledge and judgment the stock-owner may implicitly rely whenever such diseases appear. It would be more reasonable for government to undertake to make every one a physician and surgeon for the human race than to make every stock-owner a safe medical adviser upon the diseases of his six or eight different genera of domestic animals.

To furnish an account of the non-transmissible or sporadic diseases of animals that we see in this and in other localities would necessitate the writing of a considerable book, and I cannot do better in this respect than refer you to "The Farmer's Veterinary Adviser," which I published last year, and two copies of which will be found in the Congressional Library. It is true that these sporadic diseases are greatly increased by inattention to the laws of hygiene, but they extend no further than the stock of the individual owner, and in no sense endanger that of his neighbor, nor of the country at large.

If now we come to the contagious and communicable diseases of animals, we are confronted by an entirely different state of things. Here the existence in the utmost confines of our territory of one diseased animal, or even of its dried or otherwise virulent products, is a source of danger to the entire country. Here the individual owner can plead no inherent right to preserve and treat the diseased animal at the expense of an unlimited increase of the poison with each day of such preservation, and of an imminent and ever-increasing danger to the live stock of his neighbors and of the commonwealth. If a State or county harbors such a disease it cannot expect to maintain the same free and unrestricted commerce with adjacent nations as if it bore no such elements of danger. The virus of the contagious disease may be compared to a seed which in suitable soil and climate undergoes an extraordinary increase with each successive generation, and is only limited by the lack of new ground into which it may spread. Now we have at the mercy of such diseases no less than 90,000,000 head of farm quadrupeds, of a money value of nearly $200,000,000, and all this is placed in jeopardy by the existence of contagious diseases, whether generated in our own land or imported from abroad. But the money value of the whole of our live stock furnishes but an imperfect idea of the losses that would be entailed upon us consequent on the general diffusion of contagious disease. Some of the most deadly plagues, such as rinderpest, bovine lung-fever, sheep-pox, and hog-cholera, prove fatal to about one-half of the animals attacked, and as a new and susceptible generation is exposed every year, the monetary depletion in a generally infected country is to be estimated rather by the amount of yearly increase in numbers than by the losses of the first year. The results of such plagues are to be looked upon as a yearly tax of the most oppressive kind, which tend to increase in all cases, by extension, with the lapse of time, and which will always be heightened in equal ratio as we improve the kind and multiply the numbers of our live stock. What is still worse, the permanent fertility of the soil is in a great degree dependent on the numbers of the live stock which it supports, therefore any inevitable reduction of the animals, or anything that renders the soil or district inimical to such animals, will lay the foundation for an increasing sterility whenever it is unremunerative to bring manures from a distance. If we now consider that in self-supporting countries four-fifths of the population live by the cultivation of the soil or by the rearing of stock, we can estimate the stupendous interests involved in the occurrence of such pestilential devastations. As regards property at stake, we own incomparably more live stock than any nation of Europe, Russia alone excepted. The following table, giving a comparison of our live stock with that of the two foremost European nations, will amply illustrate this:

	Horses and mules	Cattle.	Sheep.	Swine.
United States (1875)	11, 149, 800	27, 870, 700	35, 935, 300	25, 726, 800
Prussia (1877)	3, 352, 237 (1867)	7, 996, 818	24, 262, 087	4, 875, 114
Great Britain and Ireland (1877)	2, 790, 851 (1874)	6, 115, 491	30, 313, 941	2, 422, 832

In absolute numbers, then, we exceed those nations by three, four, and even five times in all classes of farm-animals excepting sheep; and yet, in relation to our territory, our live stock is very deficient. We must increase our live stock if we would maintain the fertility of our land; and when our stock approaches, as it one day may, to that of the whole continent of Europe, we will be exposed to dangers equal to those of Europe in centuries past, if we continue to ignore the animal pestilences in our legislation.

As illustrating the possibilities of such losses, I may state that a single extension of such a disease as rinderpest has cost Western Europe as much as 30,000,000 head of cattle, probably worth $1,500,000,000. In eighty years of the last century it cost France alone 10,000,000 head of cattle. (Faust.) In the six years preceding 1862 lung-fever and epizootic aphtha cost Great Britain over 1,000,000 head of cattle, worth at least $50,000,000. (Fifth Report of Medical Officer of Privy Council.) In eighteen months of the prevalence of rinderpest, in 1865-'66, the same country lost about $10,000,000. It would be easy to go on at length with such statements of loss by different nations, but it will be more profitable to particularize some of the diseases that prevail among us, or threaten us.

Hog-cholera.—In the absence of reliable statistics it is impossible to estimate our yearly losses from *preventable* disease. But it is estimated that during last year one of our native-animal plagues—the so-called hog-cholera—swept off not less than $20,000,-000 worth of stock, and that one-fourth of this loss occurred in Illinois. Several destructive outbreaks occurred in this vicinity as the result of importing western hogs, and but for the comparative scarcity of hogs in this region, would have proved much more disastrous. Reports from different counties in Illinois show the present season as almost equally pestiferous, and doubtless in the absence of preventive measures similar ravages will recur at frequent intervals, whereas the merest fraction of the loss would sustain an efficient system of prevention, and leave ample margin for maintaining a veterinary college and experimental station which would be a credit and safeguard to the country. I need not say more on this affection, having recently furnished your department with an extended essay on the subject. (See Department Report, 1875.)

Texas fever.—Next to hog-cholera perhaps the disease which at present most engages the public mind is the fever produced by cattle from the Southern States mingling with our northern herds. During the great excitement of 1868 measures were adopted to prevent the introduction of such southern cattle into our Northern States, excepting during the frosts of winter. But immunity soon bred carelessness, and now the summer traffic has again acquired wide dimensions, and every year we suffer extensive losses in our Northern and Eastern States. Within the last month I have traced no less than four outbreaks in New York—at North Bangor, Franklin County; Watertown, Schenectady, and Brighton, Monroe County. These are mere straws indicating the direction of the current, and doubtless many other smaller outbreaks have occurred at other points, as they are rarely acknowledged so long as the parties interested can preserve the secret. It is only when, as at Cleveland, Ohio, the losses become so general that it is impossible to conceal them that the general public are apprised of the occurrence. The losses at North Bangor up to date have been seventeen, at Cleveland one hundred and thirty-nine. The losses in such cases, however, are not to be estimated by the deaths occurring on the infected pastures, but also by the loss of fodder incident to the disease of such pastures by the stocking of them with horses or sheep, or to the fatal results occurring at a distance to which the hay from such fields has been sent. In all the above-mentioned cases the trouble has supervened on the importation of southern cattle, and the parasites (ticks) of these are found on their northern victims.

Nothing can be simpler or more certain than the prevention of this disease, but it will never be permanently established by other authority than the general government. Safety consists in restricting the northern exodus of cattle to the winter season, and sometime before the last frosts. But it is not to be expected that the Middle States will prevent the through traffic which brings no danger to themselves, and the means can easily be found to ship and reship, so that the stock appears to come from a salubrious locality. (For description of Texas fever see department report on diseases of cattle, 1871; also report of New York board of health, 1868.)

Lung-fever.—This is a much more redoubtable affection than Texas fever, which is limited in its prevalence to our northern latitudes by the appearance of frost. Lung-fever knows no limitation by winter or summer, cold or heat, rain or drought, high or low altitude. In Western Europe and America it is a purely contagious disease, dependent alone on the pre-existing virus, and never arising spontaneously. This is amply proved not only by the records of the invasion of Ireland, England, Scotland, America, Australia, the Cape of Good Hope, Norway, Sweden, and Denmark, but also by the preservation of countries (Norway, Sweden, Denmark, Schleswig Holstein, Oldenburg, Mecklenburg, Switzerland, and the Channel Islands, Massachusetts, and Connecticut), which have treated it as an exotic, and even of such localities in plague-stricken countries as breed their own stock and never import strange animals. Of the latter may be particularly mentioned the Highlands of Scotland, certain portions of the Cheviots, and parts of Normandy. This is the most insidious of all plagues, for the poison may be retained in the system for a period of one or two months, or even more, in a latent form, and the infected animal may meanwhile be carried half way round the world in apparent health, yet bearing the seeds of this dread pestilence. And this malady we harbor on our eastern seaboard, where it is gradually but almost imperceptibly invading new territory, and preparing, when opportunity offers, to descend with devastating effect on our great stock range of the West. There is abundant evidence of the existence of this affection in Eastern New York, in New Jersey, Pennsylvania, Maryland, Delaware, Virginia, and the District of Columbia. (See government report on diseases of cattle, 1871, and many instances in current agricultural journals.) Within the past year I have advised in the case of three outbreaks, one in Eastern New York, one on Staten Island, and one in New Jersey. At present it creates little apprehension, but we are asleep over a smouldering volcano, which only wants a little more time to gather strength, when the general infection of the country will be imminent. Spreading from the port of New York, it has already gained a substantial hold upon seven different States, including the District of Columbia, and has invaded and been repeatedly expelled from two more, and it is only requisite that it should reach the sources of our stock supplies in the West to infect our railroad cars and Eastern States generally. It will create no such panic as did the Texas fever in 1868, but by its leisurely invasion of a herd, taking one victim now and another next week or next month, and by the general infection that will be established before its true nature is suspected, it will prove far more destructive in the end than would an active invasion of Texas fever, or even rinderpest. England has lost over $10,000,000 from rinderpest in the present century, but she has lost hundreds of millions from the less dreaded lung-fever. To save ourselves from similar consequences the government should see to it that this disease is arrested in its fatal course, and thoroughly eradicated from our soil. If nothing is done the time will inevitably come when we will repeat the experience of Continental Europe, Great Britain, South Africa, and Australia, when our agriculture will be crippled, and when the extinction of the plague will be a herculean, if not an impossible, task.

Aphthous fever, rinderpest, venereal disease of horses, and sheep-pox.—These are all foreign plagues, and at present happily unknown to America. All are exceedingly contagious, and, excepting the first, cause a very high mortality. The first only, the least fatal of all, and the least likely to be imported because of its short period of latency, one to two days, has ever reached America, but the contingency which brought this is even more likely to bring the others and an incomparably more terrible devastation. One (the venereal disease of horses) is almost as insidious and as long latent in the system as lung-plague, and hence as likely to reach our shores in the frequent importation of horses from the continent of Europe. That our perils from such diseases are but poorly understood even by those who ought to know them best, I infer from a recent article in one of our most popular agricultural weeklies, in which the veterinary (?) editor speaks of rinderpest and Texas fever as identical. If one who assumes the title of V. S., and is upheld by a powerful newspaper, makes such a blunder, what are we to expect of the ordinary Congressman? Unlike Texas fever, rinderpest is very highly contagious, spreads for some distance on the air, and is unaffected by any changes of climate, temperature, or management. Texan fever, as we well know, is limited to the pastures where the southern cattle have grazed, and is at once extinguished by the accession of frost. To adopt similar precautionary measures for the two would be in the highest degree impolitic and prodigal. (For aphthous fever, see my paper in Journal of New York State Agricultural Society, January, 1871. For rinderpest, see report of Cattle Plague Committee of House of Commons, England, 1866–'67.)

Indigenous animal contagia.—Among our native contagious diseases there are still four or five that demand special attention. These are, glanders and farcy, canine madness, malignant anthrax, tuberculosis, milk-sickness, and contagious foot-rot.

Glanders and farcy.—These are but one disease under different manifestations. Highly contagious and deadly not only to horses but to man, this affection is one that demands the most stringent measures for its extirpation. And yet we are doomed to see the victims of this disease freely exposed in public, kept in livery stables, watered at public drinking-troughs, worked on threshing machines which travel from farm to farm, where the diseased animals feed from the mangers and drink from the buckets of the other horses, and palmed off upon unsuspecting customers, who little know that they are purchasing a deadly poison, which may cut off their stock and themselves by a most loathsome and painful disease. And in this State of New York our only redress is by an action for damages against the vendor when the disease has wrought its dire work upon man or beast. This is truly a deep stain on our civilization. At frequent intervals over this region we find active centers of this dread disease, but are legally helpless to apply any efficient check. (See my report on glanders, in Journal of New York State Agricultural Society, July, 1869.)

Canine madness belongs to the same terrible list. Constantly fatal in its victims, human and brute, it imperatively demands such measures as will obviate its generation and prevent its communication to man where it has been developed.

Malignant anthrax, in all its numerous forms, is the third of this permanently obnoxious group. It appears in the most varied shapes, but mainly as "black quarter" and *splenic apoplexy*; attacks all animals without discrimination, and is fatal in a high degree. In man it appears as malignant pustule and intestinal mycosis, which are no less fatal than their congeries in the brute. Properly speaking, this is not a plague, but is developed only in particular localities, propagated only by direct contact, and tends to die out if removed from its native habitat. But it owns the most indestructible of all known animal poisons, and once developed is liable to be preserved in groves, soils, fodder, skins, hair, sheds, &c., for years, or even permanently. In an extended outbreak in Western New York, in which upwards of one hundred cattle and three men suffered, the grounds, previously healthy, have retained the poison for over two years, and continue to claim new victims at intervals. The hay from such infected soil will convey the disease to a distance, and hides, hair, and horns have often conveyed it to man after they had been carried half way round the world. This disease, or group of diseases, therefore, though limited in their power of extension, are suitable subjects for legislative control.

Tuberculosis.—The fourth in this homicidal list is tuberculosis or consumption. That this is communicable by inoculation, or by feeding the discharges from the softened diseased masses, no longer admits of doubt. The experiments of Klebs, Chauveau,

Gerlach, and a number of others, have supplied ample proof of this. There is even grounds for concluding that in certain cases it is conveyed in the milk, a terrible idea, considering the number of infants to whom cows' milk is the staple diet. That this disease is common in our herds is with me a matter of frequent observation, and when the diseased have been allowed to mingle with the healthy, and their discharges dropped upon the food have been consumed by others, the decimation of the herd has been common.

Milk-sickness is perhaps beyond the need of legislative control, being confined to unreclaimed localities, yet physicians in such districts claim that many fatal cases of this affection occur in distant cities as the result of eating cheese and butter made in the unhealthy regions.

Contagious foot-rot in sheep.—This is to be classed with Texas fever. It is confined to the lands on which the diseased animals have been, but its prevalence is not checked by the supervention of cold weather. At times, where the means of communication were abundant, as eight years ago in Iowa, this disease has caused a wide-spread destruction among flocks, and brought ruin on the flock-masters. The disease should therefore be checked by, at the least, an interdict on the sale of sheep from affected flocks, excepting for immediate slaughter, and to be conveyed from infected pastures in wagons.

Parasitic diseases.—Besides these there is a long list of diseases due to parasites, which it would unduly lengthen this communication to do more than glance at, but which are highly destructive to our live stock, and in some cases inimical to human life as well. There are the various forms of parasitic mange, and notably the scab in sheep, which prove the occasion of most extensive losses. There is the fluke or liver-rot, which frequently sweeps off over 3,000,000 sheep per annum from the small island of Great Britain, and is now destroying the sheep of New South Wales, where it was introduced in the bodies of German rams. I have long sought in vain for this parasite in America, but Mr. Stewart speaks of it as already common on Long Island; therefore we may look upon it as already in our midst. There are the lung-worms of cattle, horses, sheep, and pigs, which yearly cut off great numbers of our young stock, especially in the case of sheep. An Iowa flock-master writes me that his country contains 100,000 fewer sheep than it did seven years ago, though no one had found the real cause of the great mortality until he read an account of these worms which I had published in the New York Tribune. Similar accounts come from Illinois. Intestinal worms produce destructive epizootics, notably in horses, sheep, and swine, when they have been allowed to propagate foully, and the eggs getting into wells and watercourses are often carried far from the original habitat, and form centers for the determination of new outbreaks. I know of several instances in which land has had to be evacuated because of the abundance of such germs, which render it absolutely fatal to stock.

Worms in solid organs.—But it is not intestinal worms alone that demand attention. The *hydatids* of the brains of cattle and sheep, derived from a tape-worm of the dog; the *hydatids* of the liver, &c., of man and herbivora, also from a canine tape-worm; the *hydatids* or measles of pigs and occasionally of man, from a tape-worm of man; the *hydatids* or measles of calves, also from a tape-worm of man; the liver, fat, and kidney worm of swine, and the *trichina* of man and animals, should be rooted out wherever found, and none left to develop a veritable epizootic, as has occasionally happened.

In conclusion, I submit, in view of the enormous value of our live stock, and of these multiform dangers that threaten them more or less imminently, whether it is not our duty as a people to institute a system of sanitary administration for the exclusion and extinction of animal pestilences. No need of agriculture more urgently demands recognition than does that of protection from the ever-increasing danger to our live stock from fatal contagious and otherwise communicable diseases. The one dangerous feature of such diseases, *their communicability*, is the best guarantee that they can be prevented, and imposes a duty which no people nor government can ignore without proving recreant to their trust, and perpetrating a crime against the future heirs of their national inheritance.

Dr. ARTHUR V. WADGYNEAR, Castroville, Medina County, Texas, says:

Our horses are, in general, very healthy, and I have noticed only two prevalent diseases, viz: bots and distemper. The bots are produced by two different insects, *Gastrophilus* and *Chrysops metallicus*, which deposit their eggs on the hair of the horse, on the breast and fore legs mainly, and are bitten off and swallowed by the animal. They are carried into the stomach, where they remain until the following spring, when, having attained their full size as larvæ, they are carried along the intestines and evacuated. The symptoms of a horse afflicted with the bots are uneasiness and apparent pain in the bowels. The animal falls to the ground, starts up again suddenly, paws with the fore feet, and so on until exhausted. Remedies are numerous, but I have found only one which never failed. It is as follows: Mix six ounces of epsom salts with a pint of a strong decoction of worm-seed herb (*Chenopodii Mexicani*), say eight ounces of the weed to one quart of water, boiled down to a pint; then mix with this solution about four ounces of oil of turpentine; put in a quart bottle, and drench the horse well with it. At the expiration of an hour give the animal a half pint of linseed-oil, which will soon cause the expulsion of the worms or bots.

The symptoms of distemper are: Loss of appetite, swelling of the glands of the jaw and under the belly, slight fever, cough, and discharge from the nostrils. If these symptoms do not abate, emaciation, general debility, and death soon ensues. The following remedy is used: One-half ounce black sulphate of antimony, one ounce muriate ammonia, three-fourths of an ounce of saltpeter, four ounces powdered gentian-root, .and two and one-half ounces of powdered *fœnum grœcum* seed, mixed, divided into eight doses, and given three times a day. The animal must be kept in a dry, warm stable where no other horses can come in contact with him. The disease generally yields within from five to seven days.

The only disease which affects cattle to any considerable extent in Western Texas is "hollow-horn, or "horn-distemper." The cause of this disease is a "hollow stomach " and an insufficient supply of wholesome food. The symptoms are gradual decay of the pit of the horn, loss of appetite, sluggishness, swelling of the eyes and head, cold horns, urine bloody, costiveness, and swollen udder. The remedy is one-fourth pound each of powdered ginger and gentian-root, one ounce of saltpeter, and two ounces of ammonia, mixed well, and a tablespoonful given three times a day in food. If the disease is of long standing, remove the purulent matter, either by sawing off the ends of the horns or by boring them with a large gimlet. The hollow should be kept well cleaned by the injection of a solution of carbolic acid, soft-soap, and water; say one ounce of carbolic acid, four ounces of soap, and one quart of water. In the early stages the disease may be cured by a generous feeding of corn-meal and good grass, and the application of the above solution to the head and neck.

The "wolves" is a disease caused by a yellow, grayish-looking fly, of a species not known to me. It deposits its eggs under the skin above the hoof of the animal. In a few days it hatches, and the mite migrates all over the body, and finally lodges itself under the skin, where it grows and undergoes its transformation as a larva. It then bores through the skin, emerging as a perfect "heel-fly." These flies appear early in the spring, and cause the death of thousands of cattle. No remedy is known.

There is no Texas or Spanish fever among cattle in Western Texas. Ticks are plentiful, but they do no harm to native stock.

Mr. M. GILLIS, Castroville, Medina County, Texas, says:

A disease is prevailing at this time among horses called "loin-distemper," which is very fatal. The first symptoms are observed in the hind limbs. The loins seem weak, and in a few days the animal is unable to stand, its hind legs failing to support it. By many the disease is thought to be contagious, while others regard it as an affection of the kidneys. No remedy has been discovered. Colts, particularly those that come in late in the season, are attacked by ticks in such numbers that if they do not directly kill the animal they cause wounds which draw the blow-flies, which eat away the flesh and soon cause death. As a remedy for ticks, we apply coal-oil, and for maggots chrysalic ointment.

Among sheep there are several diseases, but scab causes more trouble and loss than all other diseases combined. Eighty per cent. of the flocks are afflicted with this disease. Tobacco is the common remedy, and if properly used will invariably prove beneficial. Of good tobacco twenty-five pounds to one hundred gallons of water, with a small quantity of lye or sal soda, will make a solution that will cure the disease if applied at intervals of ten days.

A disease known as "lumbers," which is a collection of worms in the stomach, some years kills almost all the lambs when but a few months old. A disease called "scours," a looseness and running off at the bowels, also proves quite fatal.

Mr. W. W. FARNSWORTH, Waterville, Lucas County, Ohio, says:

There is some complaint of hog-cholera in our vicinity, but I have not been troubled with it. We keep our hogs in clover in the summer and in clean covered pens in winter, and feed with corn-meal. We give pure water to drink. We select the best swine to breed from, and do not breed too young, as it weakens the constitution.

Mr. WILL C. RANNEY, Cape Girardeau, Cape Girardeau County, Missouri, says:

While I have no pretensions to a scientific knowledge of diseases of farm-animals or their treatment, it has been my misfortune during the last twenty years to witness much of the so-called hog-cholera, and that length of time will, I think, embrace the period of its prevalence here, as prior to that time I heard nothing of it. I have had a residence here of fifty-three years. I have known the disease to rage here among swine under all circumstances and conditions in which swine are kept. I have known it to destroy droves of hogs of the common kind, others of no breed at all—such as run in the swamps and never saw corn. I have known the same breed and quality of hogs destroyed that run in the woods around the farm with free access to running water and corn sufficient to live on. I have known them to die in cleanly pens, where they were well cared for, as rapidly as in filthy stys. I have known them to die of the disease in woods lots, where they were fed regularly and abundantly twice a day on corn. I have known them swept off by the scores in a grass pasture, while others in an adjoining clover field escaped; while at other times the clover pasture afforded no better protection than other places. Nor have those who pay the greatest attention to fine breeds of hogs and their cleanliness and comfort fared any better than those who pay no attention to either.

When the disease first made its appearance among my hogs, I was in the habit of placing the diseased animal on its back and pouring a tablespoonful of powdered copperas down its throat. Every animal thus treated at that time got well. The same treatment the past season was without any beneficial results. I know of but one animal that recovered during the past summer, and it was treated to a copious administration of pine-tar poured down its throat.

The hog-cholera presents itself under so many different phases that it would be difficult indeed to describe it. Each farmer in this vicinity would no doubt give a different description of it as it prevails among his own stock. Sometimes the animal will vomit and purge; sometimes one of these symptoms will be prominent and the other entirely lacking; sometimes neither will be observed, but the animal will apparently be affected with sore-throat. At other times sores will appear all over the body, occasionally causing the loss of the animal's eyes or ears, and not unfrequently both. Sometimes the most prominent symptom will be thumping, as a horse affected with the thumps.

I am satisfied that a knowledge of the cause or causes of the disease has not been discovered, and I can but hope that the investigations making by your department will result in finding either a preventive or a cure for this frightful and fatal disease.

Mr. J. S. O. BROOKS, Etna, Smith County, Texas, says:

No disease particularly worthy of note exists here except among hogs, which is always called cholera. I can add nothing to the statements of the learned contributor of Rhode Island, as published in Agricultural Report for 1861, as it relates to a description of the disease, its progress and various phases. Cleanliness, pure air and water, which this writer deems so important, do not appear to reduce the death-rate of our hogs in the woods.

Hundreds of remedies have enjoyed a high reputation for a time, only to be cast aside after repeated failure. The most discerning agree that nothing seems to cure. Some get well without any attention whatever.

Whether the disease is atmospheric and contagious cannot be decided—some facts point one way and just as many in another direction. It crawls slowly but surely into every nook and corner, and sometimes with very singular manifestations. The loss in 1876 was 66 per cent.

Mr. W. H. DENNY, Crockett, Houston County, Texas, says:

The domestic animals of our county, with the exception of hogs, are generally very healthy. Those that die, as a general thing, do so of old age, poverty, or accidental injuries. All hogs that die here are said to die of cholera. It matters not what the symptoms are, the duration of the disease, or anything else, whenever a hog gets sick one or more of the several remedies which are being daily published in the papers for hog-cholera are administered. It may be that the hog has quinsy, pneumonia, or enteric inflammation—it is immaterial which—he is certain to get copperas, blue-stone, sulphur, salt, soft soap, turpentine, carbolic acid, ashes, charcoal, calomel, tannin, &c. Many of the remedies used must necessarily be hurtful, and some positively destructive of life. Most of our farmers have no knowledge of the pathology of the disease of hogs or other farm-animals and are therefore not competent to give a correct diagnosis of disease; consequently the treatment of sick animals is wholly empirical, routine, and frequently destructive of life.

Mr. LAFAYETTE ROSS, Tulip, Dallas County, Arkansas, says:

My hogs were nearly all sick and about half of them died last fall. The symptoms were very different, but the results were about the same. Some would cough and linger for a week or so and then die. Others would breathe rapidly, as if greatly fatigued, while still others would be purged, &c. My neighbors generally suffered as badly or even worse than I did. Those that escaped in the fall are losing their hogs now (February 2). A great many remedies have been tried, but with little success.

Mr. CYRUS RICE, Sardinia, Erie County, New York, says:

Occasionally during the past twenty years we have lost a few cattle by a disease which I think is diphtheria. Many have, no doubt, died for a lack of a knowledge of the disease, and others because remedies were not applied soon enough. The first symptoms are profuse weeping, quick and labored breathing, driveling, and, as the disease advances, the pulse quickens. In the last stages of the disease the blood courses through the veins like a running stream. The animal refuses to either eat or drink, its flanks settle in, and it wanders around until it finally falls down and dies. After losing six head by the disease, the writer saved several others by a free use of whisky, giving saltpeter and borax in the first stages. The last-named articles (a tablespoonful of each) can be given in a bran mash once in every two or three hours, if the animal does not refuse to eat. If it refuses to take food, the throat should be well swabbed. When the disease extends up the pharynx and into the cavities of the head, and a thick, yellow matter runs from the nostrils, it is questionable if the disease can be reached so as to effect a cure. A few years since a neighbor of mine cured a cow of the disease by feeding saltpeter and borax in the inside of potatoes, which she would eat. A year thereafter the cow had a second attack, which failed to yield to treatment, and she died. I do not doubt that any medicine that is efficacious in diphtheria in a person would be good in this disease in stock, providing it was used in time. Perhaps a free use of sulphur might prove beneficial.

There is another disease that, so far as I know, has always proved fatal, although of not very frequent occurrence. It usually attacks calves, yearlings, or two-year-olds. The first symptom noticed is seen in the animal lying down, a refusal to eat, and, in a short time, inability to get upon its feet. It generally dies within from twenty-four to forty-eight hours. On taking off the hide, the legs and body, on one side, appear as if bruised to a jelly. I think the jelly appearance is the result of inflammation, but the cause is unknown here. It is sometimes called murrain, but I doubt if that is the correct name of the disease. We have no remedy.